THE MATH TUTOR'S FRIEND

BOOK 6

GEOMETRY

by
Joe Cyr

Cyr, Joseph

The Math Tutor's Friend / Book 6 – Geometry / by Joe Cyr

Third Edition ISBN: 978-1-967039-47-0 Joseph Cyr

Second Edition ISBN: 979-8-9879759-6-1
First Edition ISBN: 9798473420197 Amazon Kindle, 2022

3 4 5 6 7 8 9

Subjects: Main category of book – *geometry – lines - angles - geometric planar shapes- truth table* – high-school math

2.Other category - math tutor reference

Cover illustrations by Ramon E. Owen. Cover Design by Joe Cyr

Book design by Joe Cyr

The Math Tutor's Friend Series

Books

1 - Pre-algebra

2 - Algebra I

3 - Algebra II

4 - Algebra III

5 - Probability & Statistics

6 - Geometry

7 - Trigonometry

8 - Pre-calculus

9 - Series index & "FLASHMATH" handouts/reference cards

THE MATH TUTOR'S FRIEND

BOOK 6
GEOMETRY

Sections

About the "Math Tutor's Friend" Series:

Teachers impart knowledge guided by lesson plans in a course syllabus. Math tutors, on the other hand, face varying topics at differing levels in successive sessions; for example, pre-algebra, advanced algebra, geometry, trigonometry, statistics, or even calculus! This series addresses the issue and is intended to provide the tutor with needed quick review of high-school math topics. The material is not presented rigorously nor does it abound with proofs and explanation that are available in textbooks - the tutor already knows this stuff - but a quick refresher is always a good move before diving into a tutoring session.

Too many in high-school have reached the 11th and 12th grades lacking basic math skills such as knowing the multiplication table (relying on a calculator to multiply 6 by 7, for example), handling signed numbers, and performing operations involving fractions - knowledge that should have been acquired by the fifth grade! I try to address this deficiency by devoting the first few tutoring sessions to these fundamentals and providing handouts of multiplication tables,etc.

Books 1 through 8 of this series include a number of quiz questions (answers explained) that the math tutor may find useful.

Book 9 is an overall index of topics. It contains a set of summary "FLASHMATH" pages for quick reference or for use as handouts.

This third edition corrects typos found in previous editions and expands discussion of some of the topics.

Happy Tutoring!

Joe Cyr

Sarasota, FL

•

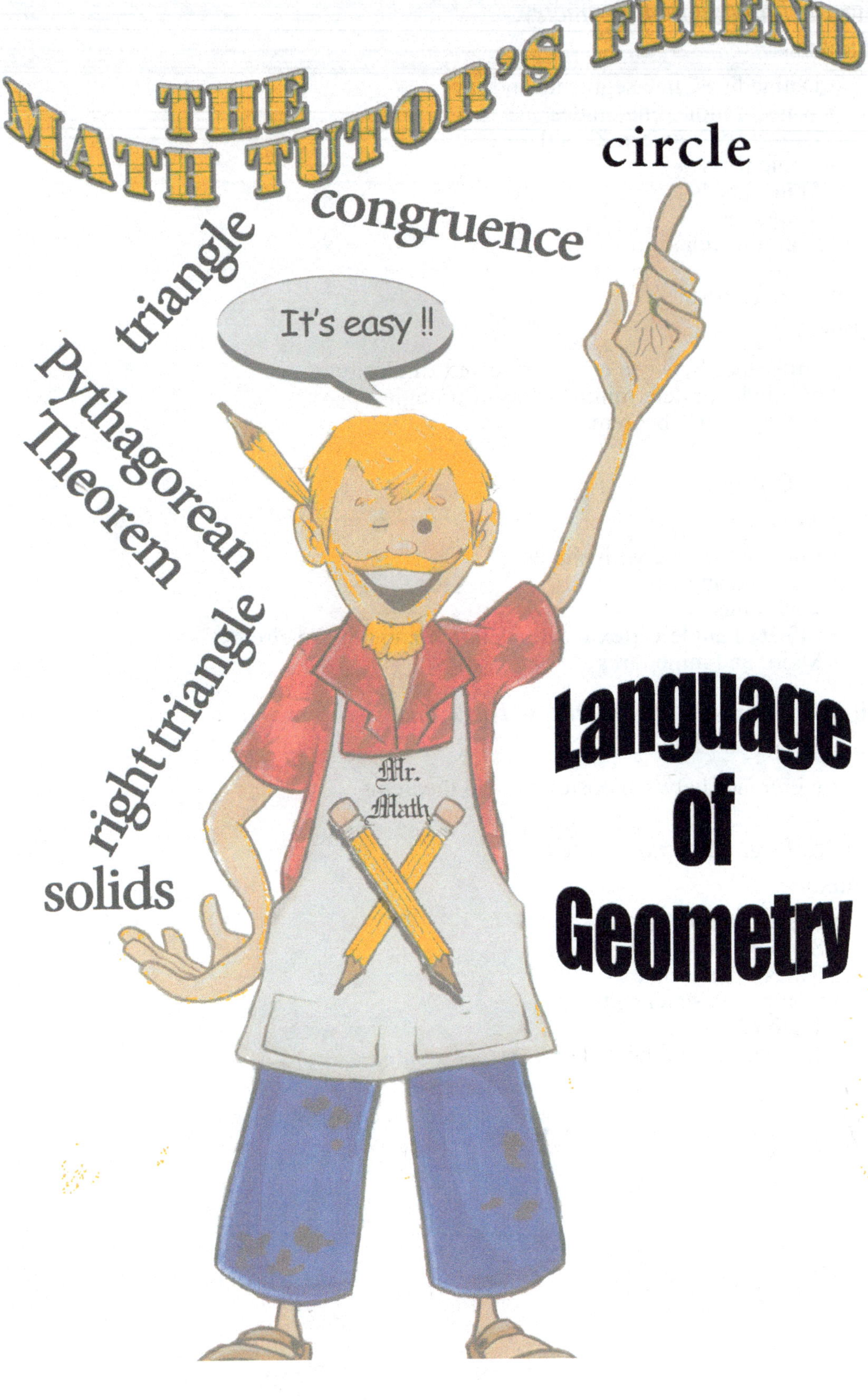
THE MATH TUTOR'S FRIEND
circle
congruence
triangle
It's easy !!
Pythagorean Theorem
right triangle
solids
Mr. Math
Language of Geometry

About Book 6: "Geometry"

Section 1, *Language of Geometry.*

Emphasis -:

- Define lines, line segments and rays
- Ways of indicating angles
- Difference between $\angle x$ and $m\angle x$
- Angle pairs
- Transversals
- Polygons
- Conic sections

Section 2, *Triangles.*

Emphasis -:

- Difference between congruence and similarity
- Methods for determining areas of triangles
- Pythagorean Theorem

Section 3, *Circles.*

Emphasis -:

- Lines associated with circles
- Arcs and angles
- Waypoints
- Effect of angle vertex locations inside and outside circles
- Major and minor arcs

Section 4, *Geometry Postulates & Theorems.*

Emphasis -:

- Difference between postulates and theorems

Section 5, *Logic & Truth Tables.*

Emphasis -:

- Logic statements
- Symbols used in logic
- Rules of inference
- Compound statements
- Truth tables
- Tautologies and fallacies

• •

1. LANGUAGE OF GEOMETRY

TABLE OF CONTENTS

page

•

1- THE LANGUAGE OF GEOMETRY

Introduction

Geometry is the branch of mathematics that involves questions of shape, size, relative position of figures, and the properties of space. It was studied by the Greeks and other ancient cultures to examine the measurement of length, area, and volume, and was instrumental in astronomy and architecture. The basic drawing tools of geometry are the *straight-edge* and *compass*, used to construct shapes, bisect line segments and angles, construct perpendicular lines, *etc*. This section addresses only the language of geometry; there are no proofs. Please refer to Section 4 for a summary of postulates and theorems and and to textbooks for proofs.

Definitions & related symbols

Points, lines, & planes

Point - a zero-dimensional location identified by the *ordered pair* (x, y) which specifies its location on the *x-y* plane of the *Cartesian Coordinate System*. Points are labeled to identify locations on a graph, e.g., $P_4(4, 1)$.

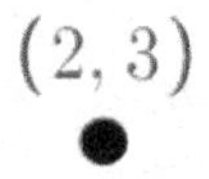

Line - a one-dimensional (length) shape of infinite length through two given points. All points on a line are *collinear.*

Line segment - a line between (and including) two different points, e.g., "line segment AB" or $\overline{AB}$.

Ray - a line with a single *endpoint.*

Opposite rays - two rays of opposite direction from a point on a line.

Labeling lines, rays, and line segments

Although a line is infinite in length, we can label it by selecting two arbitrary points on the line, for example "line AB" (see Figure 1-1 on the next page), or the entire line may be labeled; e.g., ℓ_2, $\overleftrightarrow{AB}$

A ray may be labeled by two letters, say, $\overrightarrow{AB}$. The first letter identifies the endpoint and the other an arbitrary point to indicate direction. Figure 1-1 shows *opposite rays* $\overrightarrow{AB}$ and $\overrightarrow{AC}$.

* Just as only one line can pass through two points, only one plane can pass through three points..

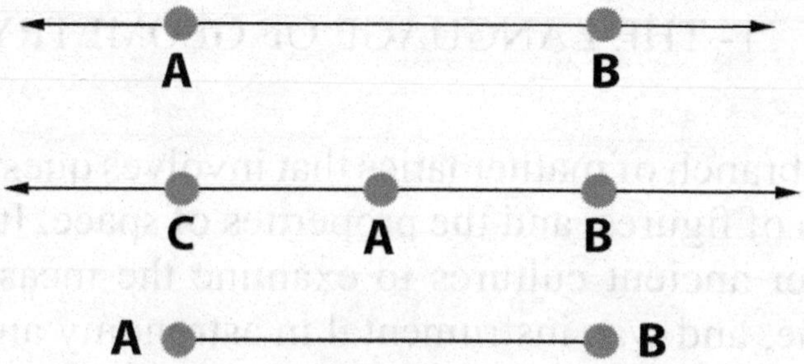

Figure 1-1

<u>*Postulate*:</u>* [line intersection]: Two lines intersect at exactly one point.

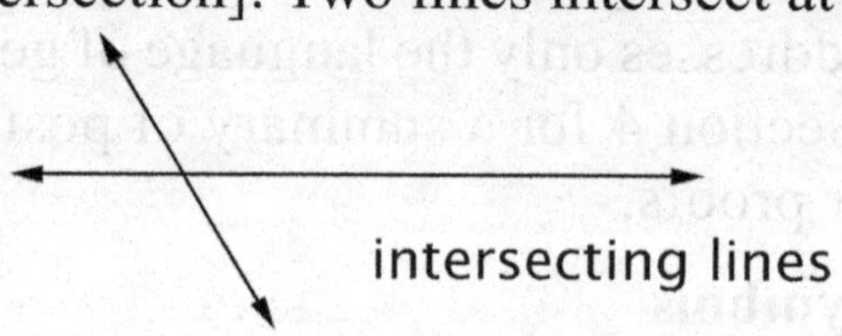

<u>*Postulate*</u> [two-point]: Through any two points there is exactly one line.

Plane [no formal definition]: - a two-dimensional (length and width) shape of infinite area defined by three points, one of which is not collinear with the other two.

<u>*Postulate*</u> [parallel lines]: Through any point not on a given line there is exactly one line that is parallel to the given line. See Figure 1-2. Identical arrows on two lines signify that they are parallel and, as we already know from algebra, have identical slopes.

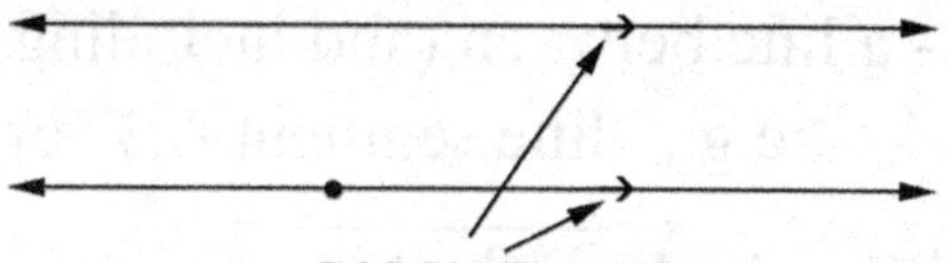

PARALLEL LINES

Figure 1-2

Angles

An *angle* is formed by two rays or line segments with a common endpoint. The rays are called *sides* of the angle and the endpoint is the *vertex*. Figure 1-3 shows an angle

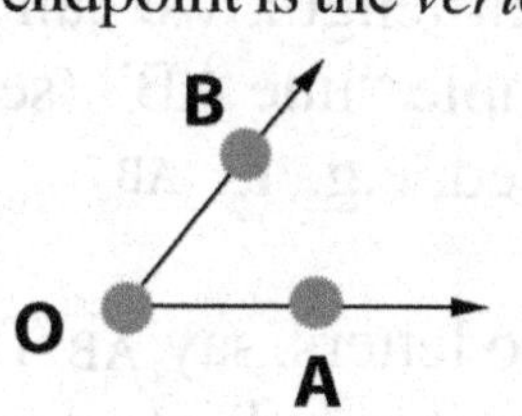

Figure 1-3

AOB with vertex *O* and sides formed by rays *OA* and *OB*. *Angle measurements* may be in *degrees*, *radians*, or less commonly, *gradians*. A complete rotation, or *turn*, is 360 degrees (360°), 2π radians, and 400 gradians (400 grads). The notation for the <u>measure of angle</u> *x* is "*m∠x*"

Notation: an angle may be referred to by a label at its vertex (e.g., "angle B), by three labels that include that of the vertex and two "pseudo-points," one on each leg (e.g., "angle *AOB*" as in Figure 1-3; the middle letter should be that of the vertex.), or by the notation "$\angle x$," which means the angle with vertex x.**

Types of angles

(Refer to Figure 1-4)

right angle - an angle of 90° (π/2 radians);

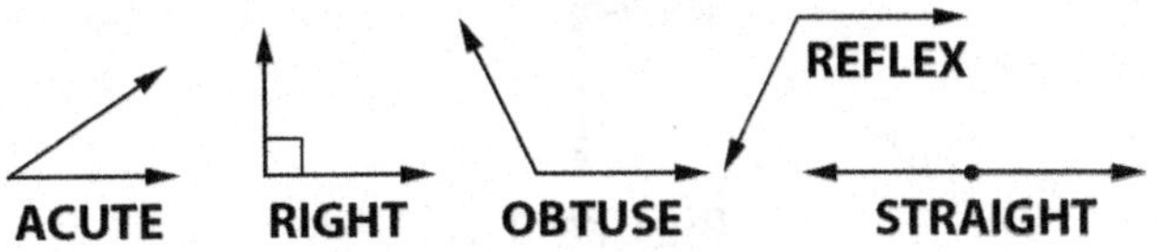

Figure 1-4

acute angle - an angle of less than 90°; (not a "pretty" angle)

obtuse angle - an angle of more than 90°;

Every acute angle is *supplementary* to an *obtuse angle* and vice-versa.

reflex angle - an angle between 180° and 360°;

straight angle - an angle of 180°, i.e., a straight line.

REMARK. Note the "square" symbol on the right angles in Figures 1-4 and 1-5; if this symbol is not shown, do not assume that the angle is a right angle, even though is might appear to be so.

Angle pairs

(Refer to Figure 1-5)

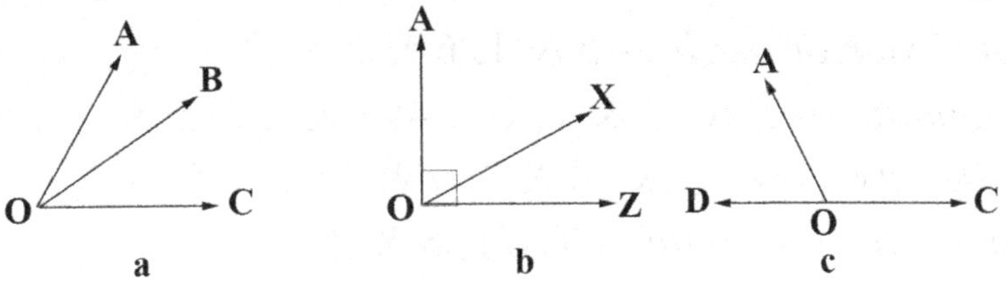

Figure 1-5

adjacent angles - two angles having a common side and a common vertex but with no common interior points;

complementary angles - two angles whose angular measurements sum to 90° (angles need not be adjacent);

external angle - Angle formed by extending the side of a polygon.

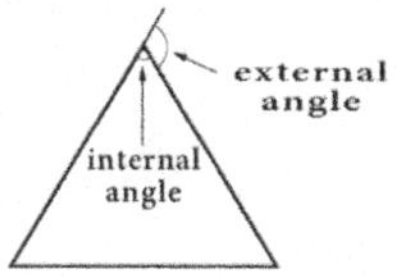

linear pair (also straight angle) is formed by opposite rays.

supplementary angles - two angles whose angular measurements add to 180° (angles need not be adjacent); if adjacent, the angles are a linear pair.

vertical angles - Two angles whose sides form two pairs of opposite rays. Figure 1-6 shows many pairs of adjacent angles.

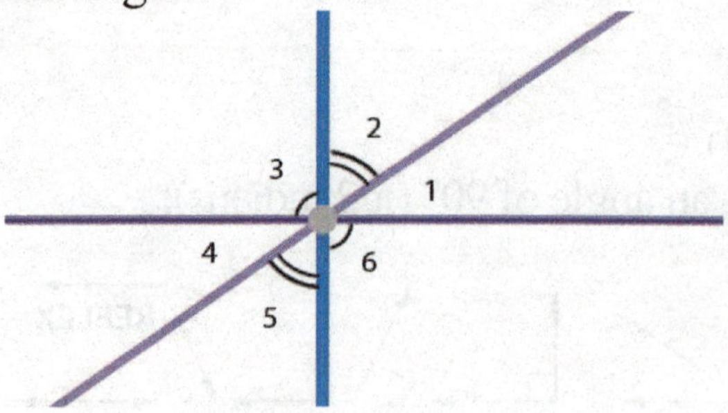

Figure 1-6

Angle 1 is adjacent to Angles 2 and 6, for example. Angles 1 and 2 are complementary. There are no supplementary adjacent angles in the figure. There are three pairs of vertical angles: ∠1 and ∠4; ∠2 and ∠5; ∠3 and ∠6.

Transversals

Transversal - a line that intersects two or more lines at different points.

Angle pairs associated with the transversal

Figure 1-7 shows a transversal crossing a pair of parallel lines (how do you know that the lines are indeed parallel?) Note the angles that are formed.* One can readily identify the vertical angles and supplementary adjacent angles at each intersection, but there are additional angle pairs of interest:

alternate interior angles - 3 & 7; 4 & 8

alternate exterior angles - 5 & 1; 6 & 2

corresponding angles - 1 & 7; 2 & 4; 3 & 5; 8 & 6

same side interior angles - 3 & 4; 7 & 8

same side exterior angles - 6 & 1; 5 & 2

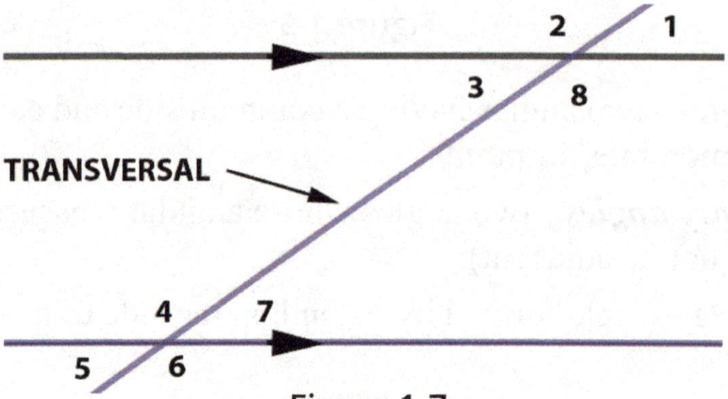

Figure 1-7

* A complete understanding of transversals and associated angles is essential in geometry.

Two-dimensional geometric shapes

Circle - the infinite set of points at a fixed distance called the radius, r, from a given point called the center.

diameter - distance between *antipodes* of a circle. The diameter, d, is twice the radius: $d = 2r$. The vertical line passing through O in Figure 1-8 is the diameter.

circumference, ***C*** - length of the outside of the circle: $C = 2\pi r$; think of the circumference as the perimeter ("walk around") of the circle.

arc - a continuous portion of the circumference. The length of an arc, **s**, is $s = r\,\theta$, where θ is a central angle measured in radians.

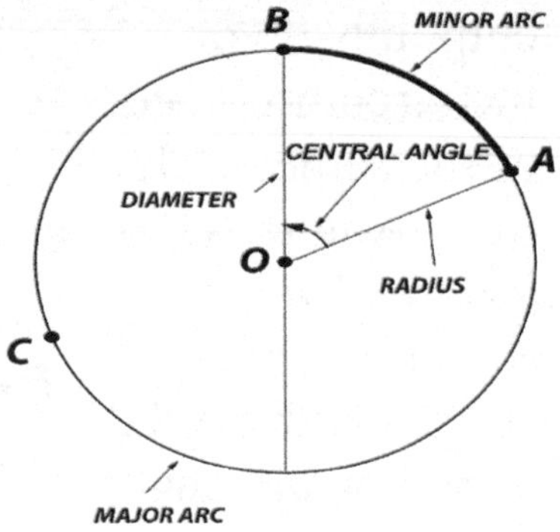

Figure 1-8

central angle - an angle formed by two radii of a circle; that is, the vertex of a central angle is the center of the circle. See Figure 1-8. Angle AOB is a central angle. The part of the circumference from A to B is the *minor arc*, and the part from B through C to A is the *major arc*. The sum of the minor and major arcs in this example is equal to the circumference of the circle. A minor arc is less than 180° and the major arc is greater than 180°.

semicircle - The arcs on either side of the diameter of a circle form *semicircles*, which are half-circles (exactly 180°).

Lines associated with circles

chord - a line segment with end points on the circle. See Figure 1-9.

tangent - a line or ray that intersects ("kisses") the circle at exactly one point of the circumference;

secant - a line or ray that contains a chord of a circle;

inscribed polygon - a polygon inside a circle whose circumference contains all of the vertices of the polygon;

circumscribed polygon - a polygon whose sides are tangent to the circle; that is, the circle is inside the circumscribed polygon.

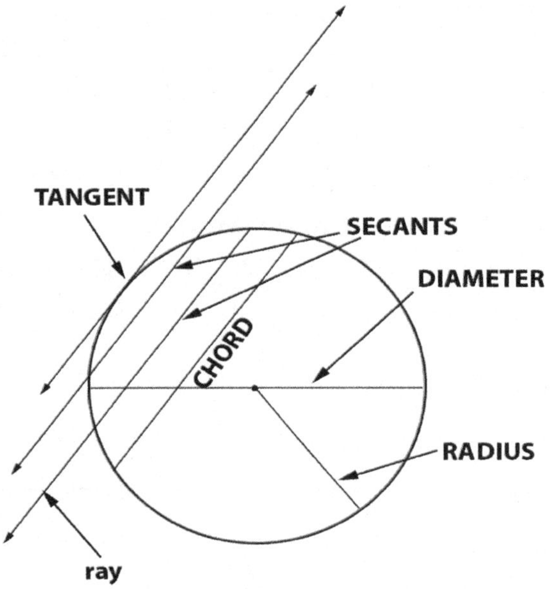

Lines associated with circles

Figure 1-9

Angles associated with circles
(refer to Figure 1-10)
central angle - an angle whose vertex is the center of the circle and whose entpoints lie on the circumference of the same circle. (e.g., Angle AOG)
inscribed angle - an angle formed by chords of a circle; the vertex of the angle and the endpoints of the sides lie on the circumference (e.g., Angle BDF);

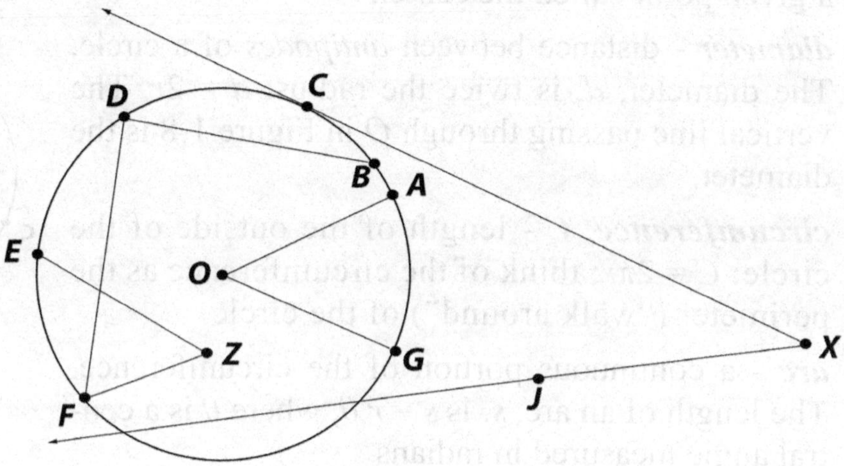

Figure 1-10

internal angle - an angle whose vertex is within the circle, as with the point of intersection of two chords (e.g., Angle FZE, not formed with chords);
Note that both central and internal angles partition the circumference into two arcs, the smaller is the *minor arc*, the larger the *major arc*;
external angle - relative to circles, an angle with vertex outside the circle and whose sides intercept the circle or are tangent to the circle. In Figure 1-10 the angle of rays XC and XJ form an external angle. Note that ray XC is a tangent and ray XJ is a secant.

Similarity
Two polygons are said to be *similar* if the corresponding angles of both shapes are congruent and the sides differ by a constant ratio. All circles (and spheres) of any size are similar as they differ only in radii. Similarity is denoted by the symbol "~"; e.g., "ABC ~ DEF" for triangles, say.

Convex vs concave shapes
If no line can intersect the sides or vertices at more than two points, then the shape is *convex*. Figure 1-11 shows a convex shape (left); the other is *concave*.

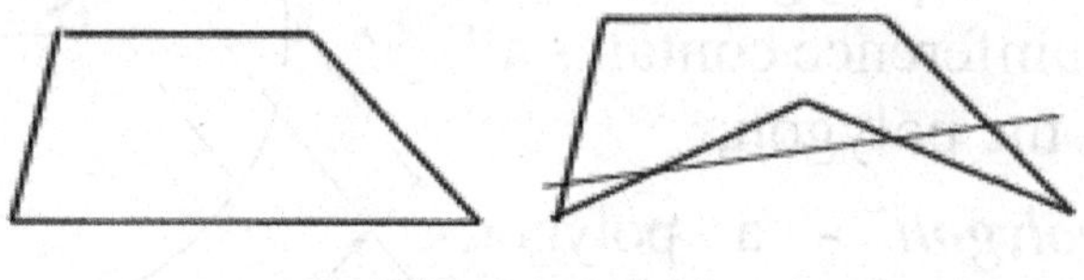
Figure 1-11

Other geometric shapes

kite - a quadrilateral having two pairs of congruent sides where opposite sides are not congruent. The long diagonal bisects the short diagonal in a kite.

cyclic quadrilateral - A quadrilateral that can be circumscribed by a circle so that the circle touches each vertex. The two quadrilaterals on top in Figure 1-12 are cyclic quadrilaterals.

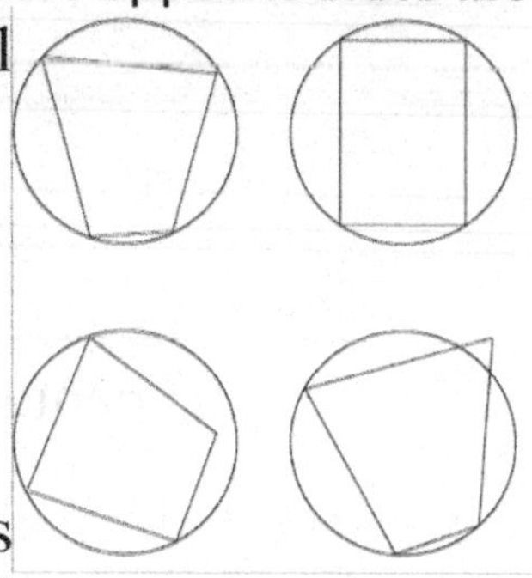

Figure 1-12

diagonal - a line segment, not an edge, joining two vertices of a polygon; this definition excludes triangles because all vertex pairs are connected by sides.

Quadrilaterals

("four sides" See Figure 1-13)

parallelogram - a 2-dimensional quadrilateral for which opposite sides are parallel.

rectangle - a quadrilateral with four right angles.

square - a rectangle with four sides of equal length.

rhombus - a quadrilateral with four congruent sides.

trapezoid - a quadrilateral with exactly one pair of parallel sides, called bases. The remaining two sides are legs.

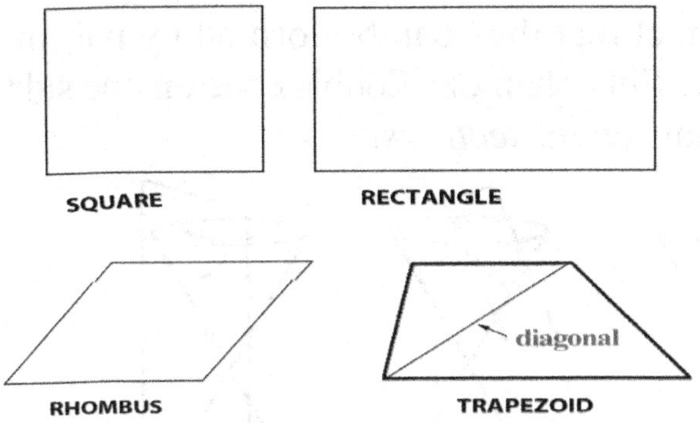

Figure 1-13

isosceles trapezoid - a trapezoid whose legs form congruent angles with the bases.

Polygons

A *polygon* is a plane figure that is described by a finite number of straight line segments connected to form a closed figure (*polygonal chain;* also *polygonal circuit*).

Regular polygons have congruent internal angles and congruent sides. See Figure 1-14. External angles of a polygon (formed by a side and extension of an adjacent side) are supplementary to the adjacent internal angle.

The sum of the measures of all of the external angles of a polygon is 360°. The measure of an internal angle of a regular polygon of n sides is $(n-2)\times 180°$.

pentagon - a polygon with five sides
hexagon - a polygon with six sides
n-gon - a polygon with *n* sides.

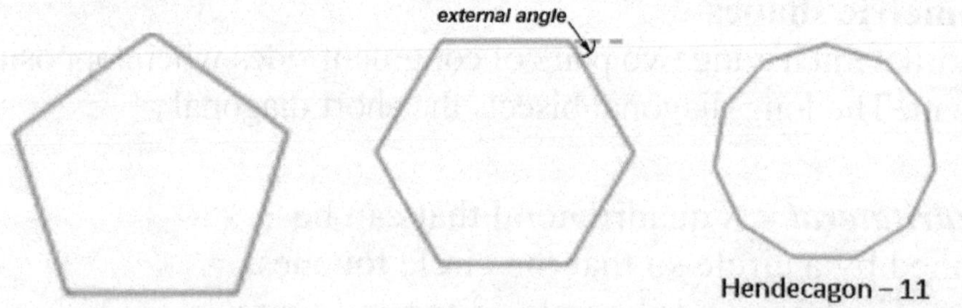

Figure 1-14

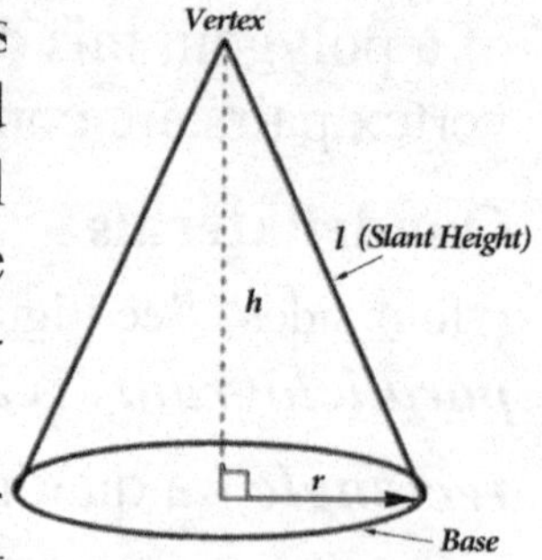

Figure 1-15

cone - A three-dimensional geometric shape that tapers s base to a point called the *apex* or *vertex*. A cone is formed by a set of line segments or lines connecting the apex, to all of the points on a base in a plane that does not contain the apex. The boundary formed by these lines is called the *lateral surface* or *nappe*.

right circular cone A right circular cone has its axis perpendicular to the plane of the base. We can generate a right circular cone by revolving a right triangle about one of its legs. In Figure 1-15, you can see a right circular cone, which has a circular base of radius ***r*** .

Conic sections

The conics get their name from the fact that they can be formed by passing a plane through a *double-napped cone*. Either half of a double cone on one side of the apex is called a *nappe*. Here are four* *conic sections*.

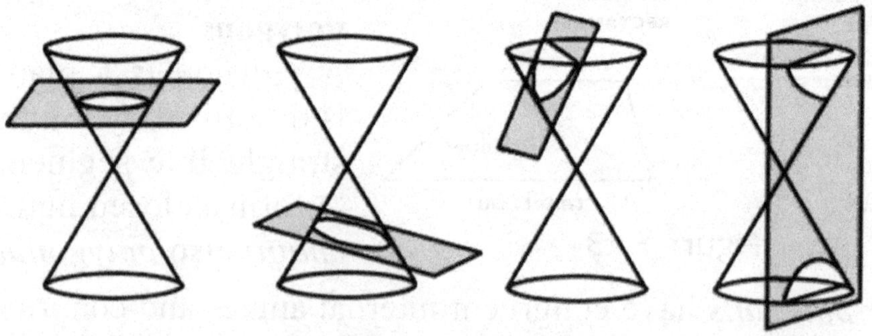

Figure 1-16

circle - formed when a plane parallel to the base intersects a cone.
parabola - formed when a plane intersects both the cone and its base;
ellipse - formed when a plane intersects a cone at an angle not parallel to the base and does not intersect the base;

* The line, ray, and point may be considered degenerate forms of conic sections; for example a plane passing between the cones at the apexes forms a point.

hyperbola - formed by a plane parallel to the axis of a double cone intersecting both nappes.

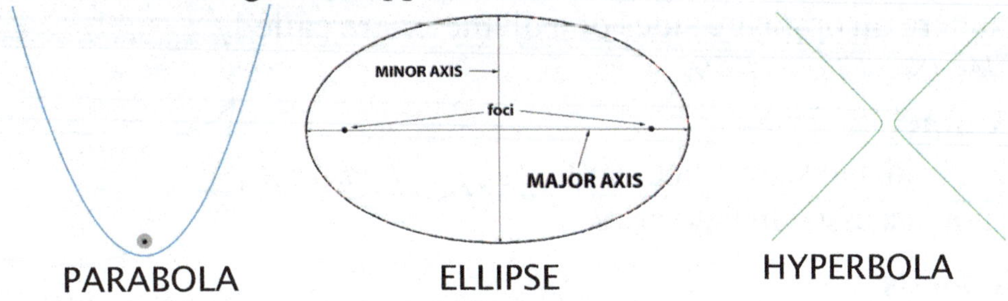

Figure 1-17

Solids

Polyhedron - a solid described by vertices, edges, and faces of two-dimensional polygons. See Figure 1-18.

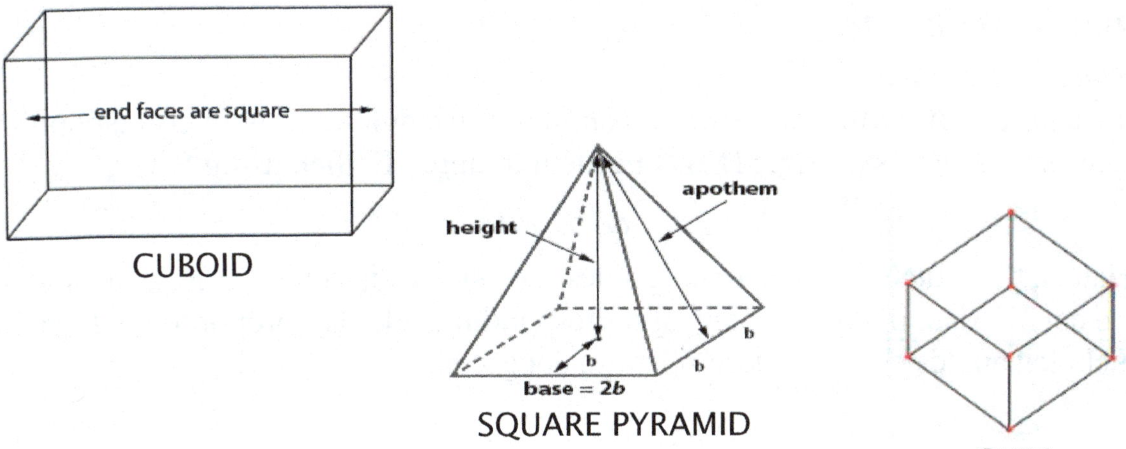

Figure 1-18

parallelepiped ("box) - a polyhedron of six faces that are parallelograms.

cube - a parallelepiped of which each face is a square.

cuboid - a parallelepiped of two square faces; the remaining faces are rectangles.

pyramid - a solid with a square base and four triangular faces.

apothem - the height of the triangle formed by a face of the pyramid.

base - the length of a side of the square base.

height (*altitude*) - the distance between the apex and base o

Cylinder - a solid geometric figure with straight parallel sid section (the latter is an *elliptical cylinder*). If the height is perpendicular to the base, the cylinder is a *right cylinder*; otherwise is an *oblique cylinder.*

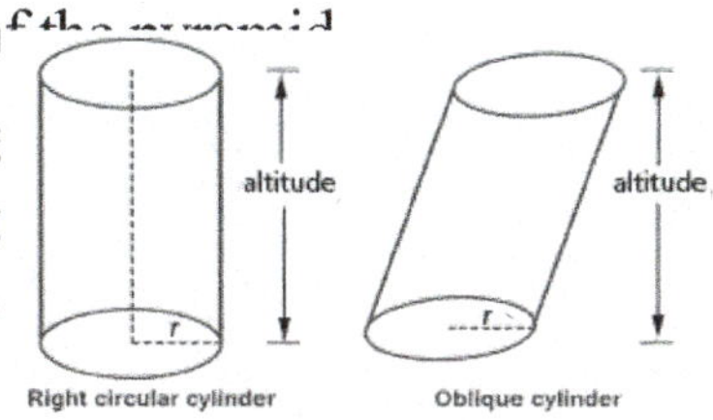

CYLINDERS

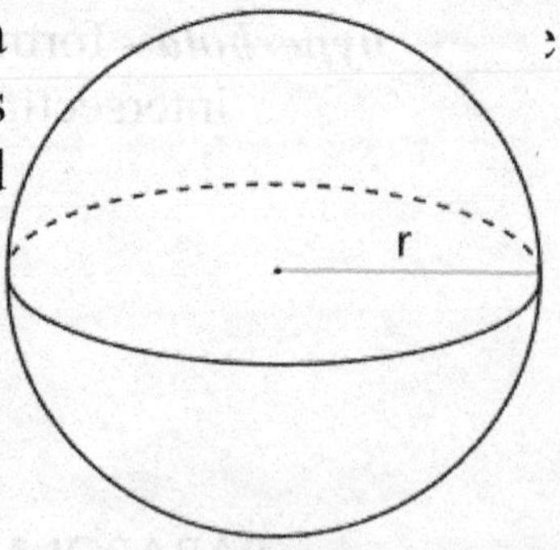

SPHERE

Figure 1-19

Sphere - a solid for which the set of all points are loca *radius*) from a given point (the *center*). Pairs of points on the sphere on opposite sides of a diameter are called *antipodes (See Figure 1-19).*

Other solids

Other solids include asymmetrical forms, *polyhedra, etc,* and are not discussed in this series.

Similar solids

Two solids with ratios of corresponding linear measures, such as heights or radii, are *similar solids*. This ratio is the *scale factor* of one solid to the other solid. If two similar solids have a scale factor of $a : b$, then the corresponding areas (e.g., lateral areas, base areas, surface areas) have a ratio of $a^2 : b^2$ and the volumes have a ratio of $a^3 : b^3$.

Ratio of similitude

(similar triangles)

If triangle ABC with sides *a, b, c,* is similar to triangle DEF with corresponding sides *d, e, f,* and say angle B is congruent to angle E, then using SAS*,

$$c/f = a/d$$

That is, the ratios of corresponding sides (i.e., sides adjacent to congruent angles) are equal. Also, if any two pairs of corresponding sides are proportional, then the included angles between these sides are congruent.

•

* Side-Angle-Side; there are also SSS, SSA, AAA, etc., used in proving congruence and similarity.

THE MATH TUTOR'S FRIEND
centroid
scalene triangle
congruence
It's easy !!
right triangle
supplement
rays
Mr. Math
TRIANGLES

2. TRIANGLES

TABLE OF CONTENTS

page

2. TRIANGLES

Triangles (“three angles”)

A ***triangle*** is a convex 2-dimensional shape consisting of exactly three sides and three (internal) angles that meet at three non-collinear points called *vertices*. The sum of the internal angles in a triangle is 180°. An *included side* is the side between the vertices of two given angles, and an *included angle* is an angle between two adjacent sides of a triangle.

equilateral triangle - all three sides are congruent;* therefore the angles are congruent and measure 60°;

isosceles triangle - two sides are congruent; the third side is the base and the angles adjacent to the base are congruent.

scalene triangle - no sides are congruent (therefore, no angles are congruent);

right triangle - see Figure 2-1; one of the angles is a right angle (90°). The longest side is the hypotenuse; the other two sides are *legs*.

acute triangle - all internal angles are less than 90°.

obtuse triangle - one angle is greater than 90°.

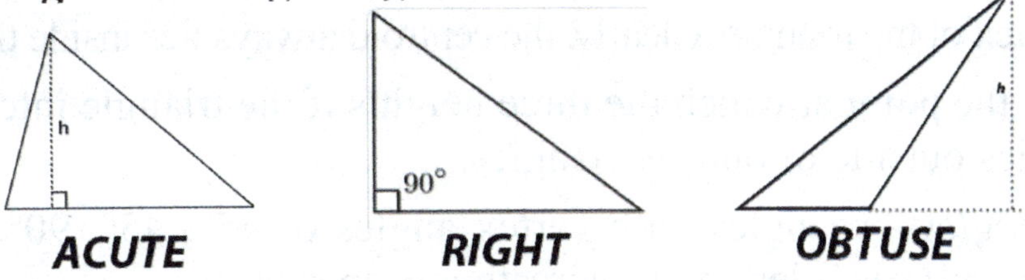

Figure 2-1

The ***height*** (symbol *h* in the figure), or ***altitude*** (*a*), of a triangle is the distance from one vertex to the side opposite (or its extension, as shown in Figure 2-1). There are, therefore, three heights. In a right triangle two of the heights are its legs. Figure 2-1 shows an acute triangle, a right triangle, and an obtuse triangle.

circumcenter (see Figure 2-2) - the point of intersection of the perpendicular bisectors of a triangle’s sides. It is the center of the triangle’s *circumcircle* (passes through all the vertices). The circumcenter can lie outside the triangle.

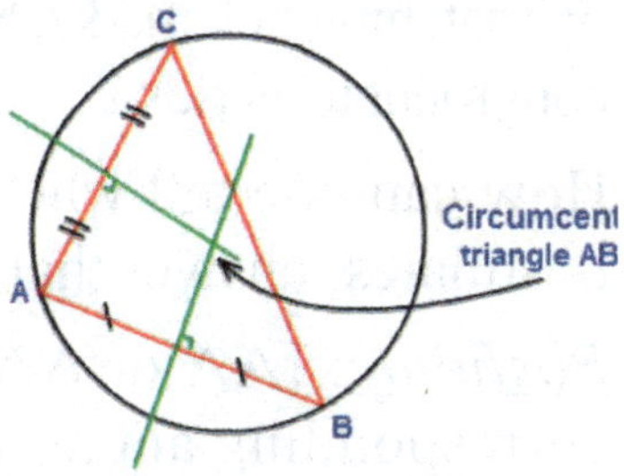

Figure 2-2

incenter - the point where the angle’s bisectors intersect. It always lies inside the triangle and is the center of the triangle’s *incircle*, the largest circle that will fit inside the triangle; the incircle is tangent to all three sides of the triangle. See Figure 2-3.

* “Congruent” means “identical.”

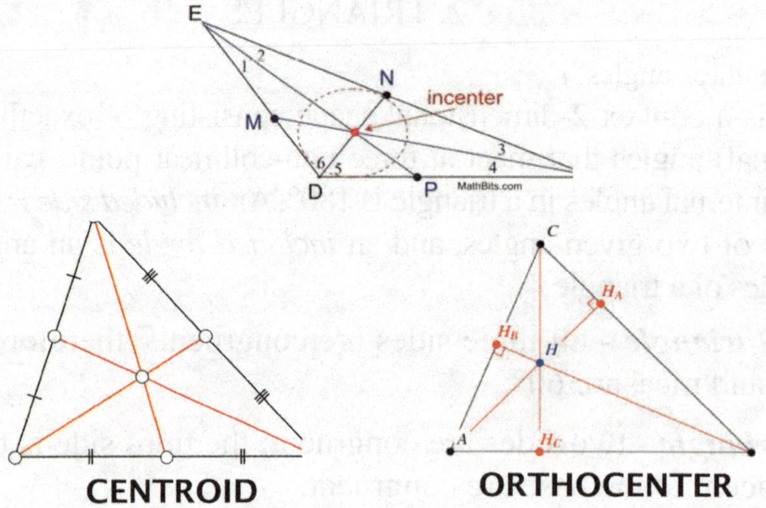

Figure 2-3

centroid - the point of intersection at of lines connecting the midpoint of a side to its opposite vertex. The centroid of a triangle is the center of mass (center of gravity) of the triangle. Cut a triangle out of cardboard or sheet metal and you can balance it on a stick or nail supporting it at the centroid. An interesting fact is that the centroid divides each median at the ratio of 2:1; that is, the centroid lies ⅔ of the distance from each vertex of the triangle. Clearly, the centroid always lies inside the triangle.

orthocenter - the point at which the three heights if the triangle intersect. The orthocenter lies outside of obtuse triangles.

standard triangles - triangles with vertex angles of 45°, 45°, 90°, and 30°, 60°, 90°. Standard triangles have interesting properties.

Congruence & similarity

Illustrating congruence. Matching sets of "hash marks" or *tick marks* are used to show that line segments are the same length (congruent). You can use just one tick mark, The sets may contain one, two or three or more tick marks - matching sets indicate congruent line segments such as triangle sides as shown below. Similarly, matching sets of *hash arcs* are used to show that angles are congruent. See Figure 2-4 on the next page. Line segment AB is congruent to CD, and EF is congruent to GH as shown by matching tick marks. Similarly, Vertex X is congruent to Vertex Y, and Vertex W is congruent to Vertex Z.

How can one tell whether triangles are congruent or similar? The following postulates answer that question (Figure 2-5):

<u>Postulate</u>: [AA or AAA] (*angle-angle*). Two triangles are similar if two corresponding angles are congruent. The third angle will also be congruent (internal angles in triangles sum to 180°) so the triangles are similar; that is the lengths of the legs of one differs from that of the other by a constant factor (twice as long, *etc.*)

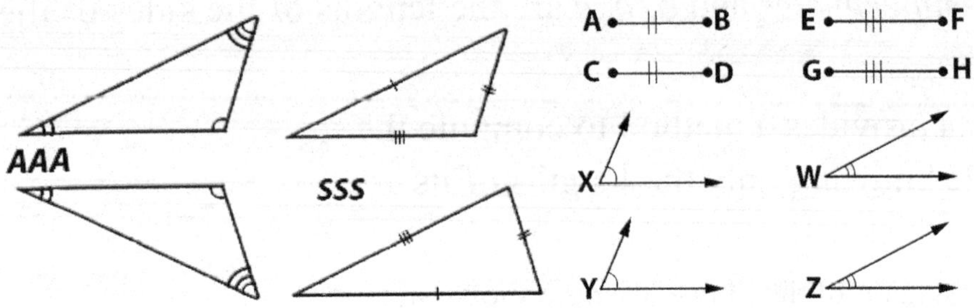

Figure 2-4

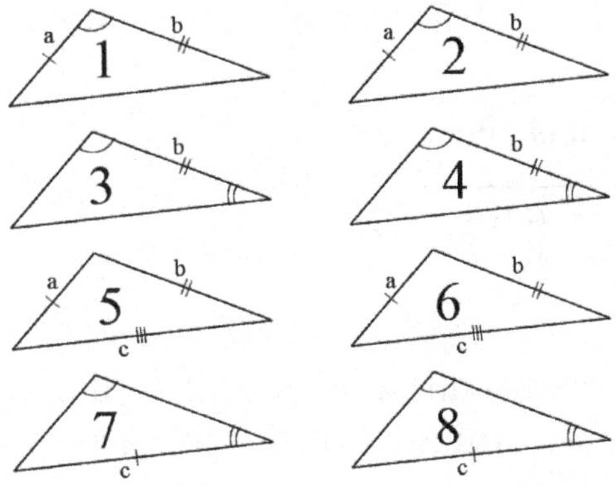

$\Delta 1 \cong \Delta 2$ by SAS

$\Delta 3 \cong \Delta 4$ by ASA

$\Delta 5 \cong \Delta 6$ by SSS

$\Delta 7 \cong \Delta 8$ by AAS

Figure 2-5

<u>*Postulate*</u>: [SSS] (*side-side-side*). If three sides of one triangle are congruent to three sides of another triangle, then the triangles are congruent.

<u>*Postulate*</u>: [SAS] (*side-angle-side*). If two sides and the included angle of a triangle are congruent to two sides and the included angle of another, then the triangles are congruent.

<u>*Postulate*</u>: [ASA] (*angle-side-angle*). If two angles and the included side of a triangle are congruent to two angles and the included side of another triangle, then the triangles are congruent.

<u>*Postulate*</u>: [AAS] (*angle-angle-side*). If two angles and the non-included side of a triangle are congruent to two angles and the non-included side of another triangle, then the triangles are congruent.

Area of a triangle

Two ways for finding the area of a triangle are *area* = A = ½ *height* × *base*, where the height is measured from one side of the triangle (the base) to the opposite vertex, and *Herron's formula*:

$$A = \sqrt{s(s-a)(s-b)(s-c)}$$

where

$$s = \frac{a+b+c}{2}$$

Here, s is the *semiperimeter* and *a, b, c* are the lengths of the sides of the triangle.

Herron's formula provides a method to compute the area of a triangle knowing only the lengths of its sides.

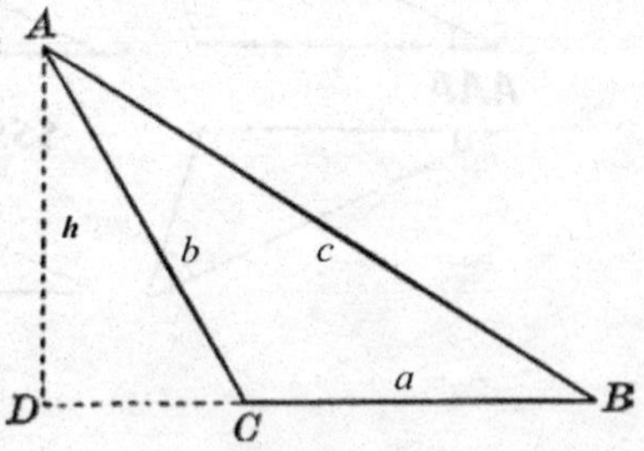

Figure 2-6

Example: The perimeter of the obtuse triangle shown in Figure 2-6 is $a + b + c$. Half of that, the semiperimeter, is

$$s = \frac{a + b + c}{2}$$

and so, using Herron's formula, the area is given by

$$A = \sqrt{s(s-a)(s-b)(s-c)}$$

Pythagorean Theorem

"In a right triangle with *hypotenuse* c and sides *a* and *b*, the square of the hypotenuse is equal to the sum of the squares of the other two sides." That is, $c^2 = a^2 + b^2$.

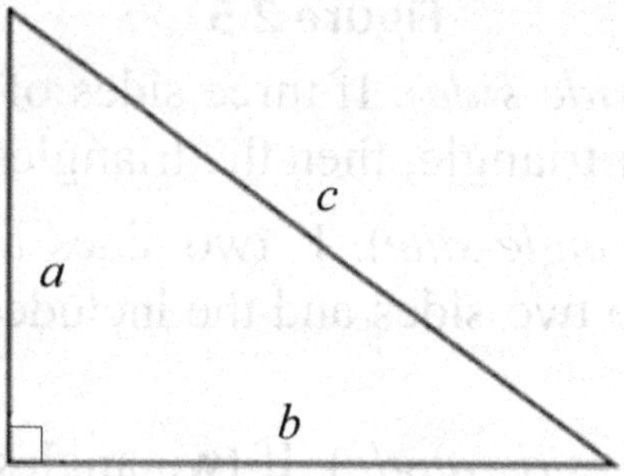

The *Pythagorean Theorem* was first proved more than 3,000 years ago. It has many proofs, including one published by an American who became president (James Garfield).

The Pythagorean Theorem is a special case of the *Law of Cosines* that will be addressed in Book 7 of this series.

THE MATH TUTOR'S FRIEND
tangent
conic section
C = 2πr = 2d
It's easy !!
chord
secant
Mr. Math
A=πr²
CIRCLES

3. CIRCLES

TABLE OF CONTENTS

page

3. CIRCLES

Circle

A *circle* is the set of points in a plane at a specified distance from a given point on the plane called the *center*. The specified distance is the *radius, r*; the set of points form the *circumference*. The distance between opposite points on the circle is the *diameter, d*. See Section 1 for a discussion of circle nomenclature and lines and angles associated with circles.

Tangent circles

Figure 3-1 shows two smaller circles *tangent* to a larger circle. The points of tangency are A and B, respectively. One of the small circles is an *external tangent*, the other an *internal tangent*. Special cases of tangents are discussed in another part of this section.

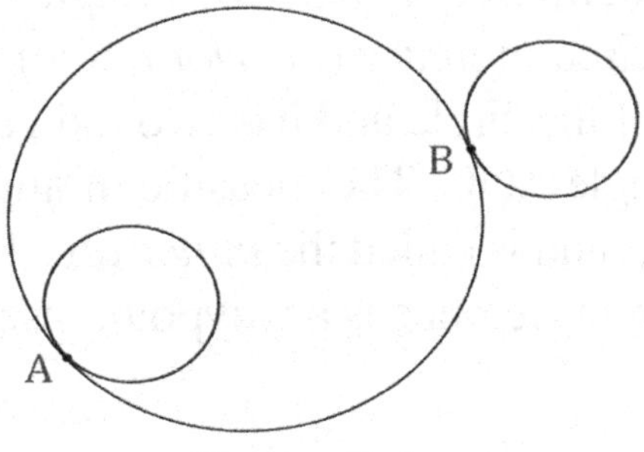

Figure 3-1

Concentric circles

Concentric circles share the same center. Figure 3-2 shows two concentric circles with a third circle intersecting the larger and tangent to the smaller concentric circle.

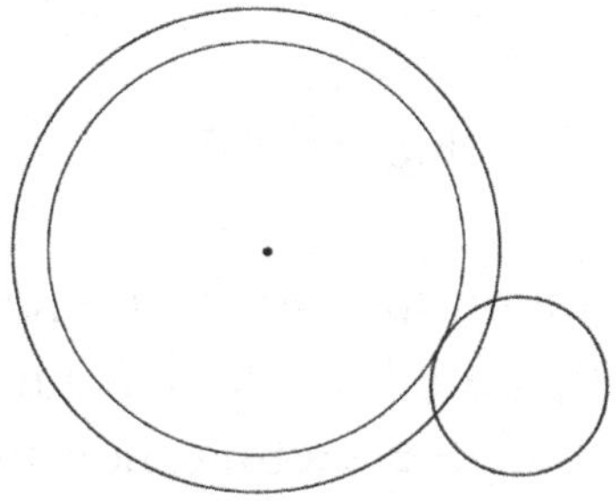

Figure 3-2

Arcs and angles*

Identifying arcs

Figure 3-3 shows a circle and two tangent rays, XA and XB, that partition the circle circumference into arc AB, the minor arc, and arc ACB, the major arc. Note that the *waypoint* C is used to distinguish one arc from the other.

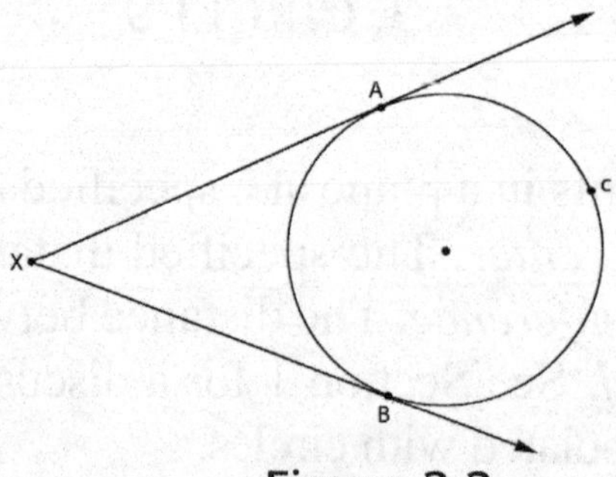

Figure 3-3

Positions of angle vertex with respect to a given circle

Central angle

A *central angle* has its vertex at the center of the circle. Vertices may be located outside the circle, as shown in Figure 3-3, or at the center (central angle) as shown in Figure 3-4, or on the circumference (*inscribed angle*) as seen in Figure 3-5, or between the center and the circumference (*interior angle*) as in Figure 3-6. An angle whose vertex is the center of the circle and has two radii as arms is a central angle. Figure 3-4 shows central angle AOC. The measure of angle AOC is denoted by m $\angle$AOC is the shaded arc AC and is called the *minor arc*. The *major arc* is arc ABC, where the point B on the circumference is a waypoint inserted to avoid confusion.

Inscribed angle

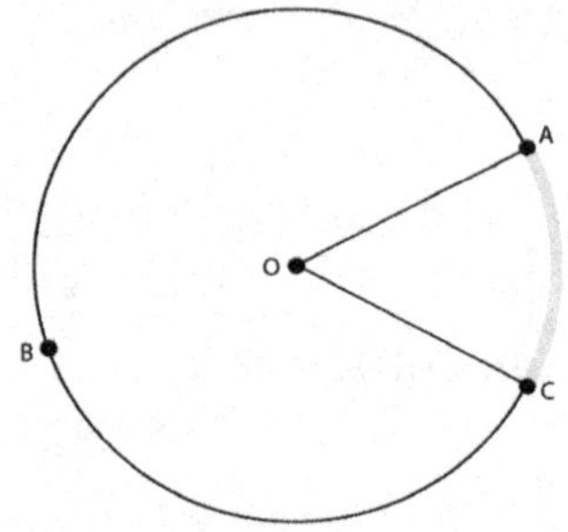

Figure 3-4

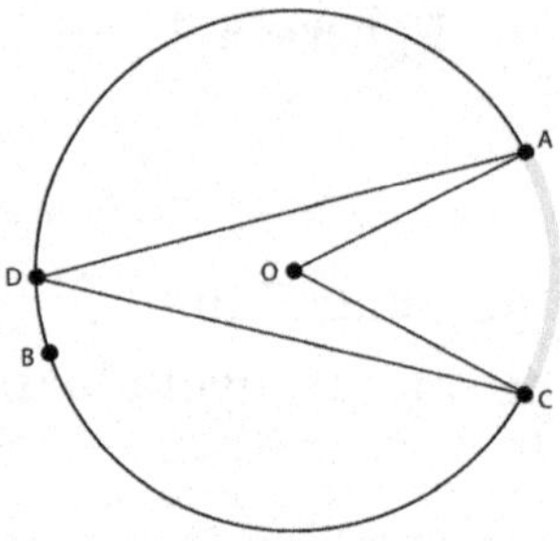

Figure 3-5

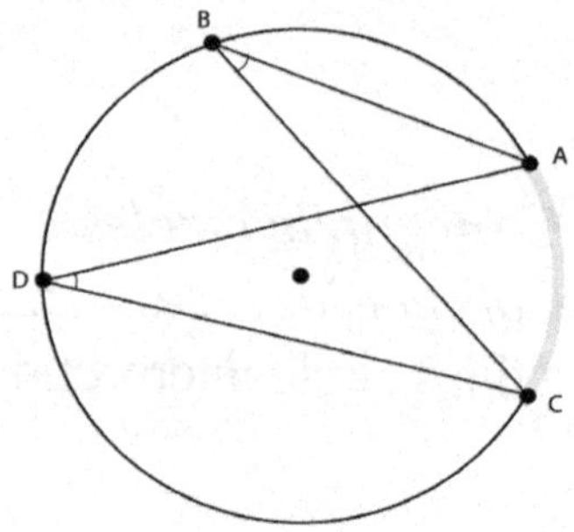

Figure 3-6

An angle whose vertex is a point on the circumference of a circle is an inscribed angle (Figure 3-5). The inscribed angle ADC has an angle measurement that is half that of the central angle AOC intersecting the same minor arc, shown in the figure by a shaded arc. Inscribed angles that intercept the same arc of a circle are congruent; see Figure 3-6. Inscribed angle ABC in the figure, is congruent to inscribed angle ADC because they intercept the same arc.

An angle can have one leg tangent to a circle and the other a secant as shown in Figure 3-7 on the next page. Rays XB and XA have common origin X and form an exterior angle with the rays as legs. Ray XB is tangent to the circle at B and ray XA intersects the circle forming chord CA.

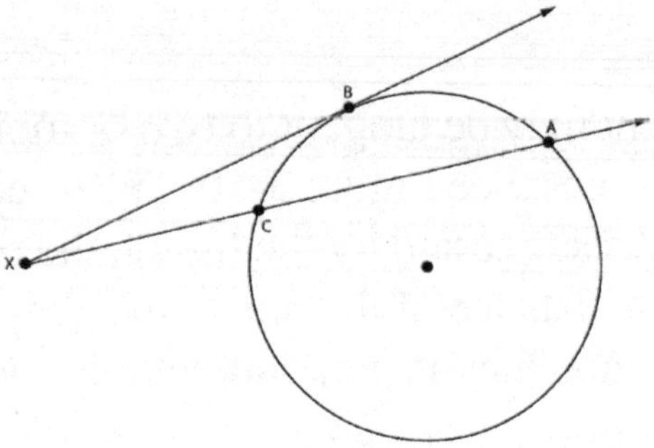

Figure 3-7

Interior angle

An angle with its vertex located inside a circle, but not at the center of the circle, is an interior angle. See Figure 3-8. Here, lines BE and CA intersect inside the circle at F which is the vertex of four adjacent angles with two pairs of vertical angles. Thus ∠ BFA is congruent to ∠ CFE and ∠ BFC is congruent to ∠ AFE.

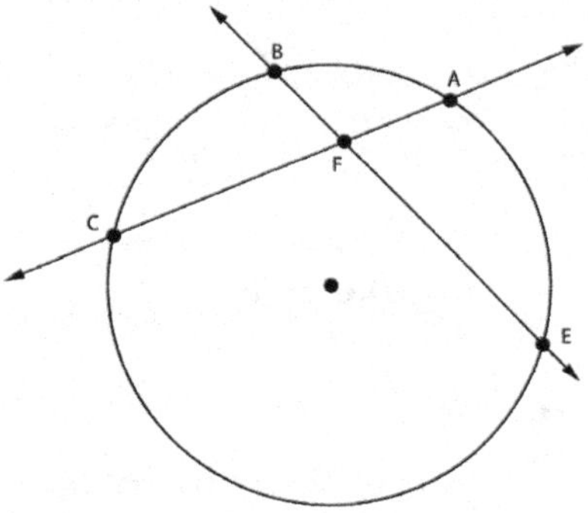

Figure 3-8

Congruent chords

Congruent chords, that is, chords of equal length in a circle or in congruent circles, are equidistant from the circle's center(s). See Figure 3-9. The lengths of chords BA and CD are equal and so are equidistant from the center of the circle.

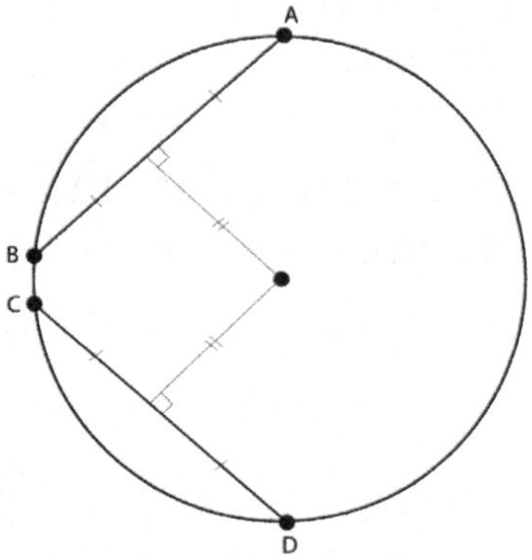

Figure 3-9

Additionally, two radii that intersect the chords will bisect the chords, as shown in the figure.

Tangents to circles

Common tangents

A line, ray, or line segment may be tangent to two or more circles. Such a line is a *common tangent* to the circles. See Figure 3-10. The three lines show are common tangents to both circles. As suggested by the figure, the point of tangency of a line and a circle is the shortest distance of the line to the center of the circle; therefore, the radius to the point of tangency is perpendicular to the tangent line.

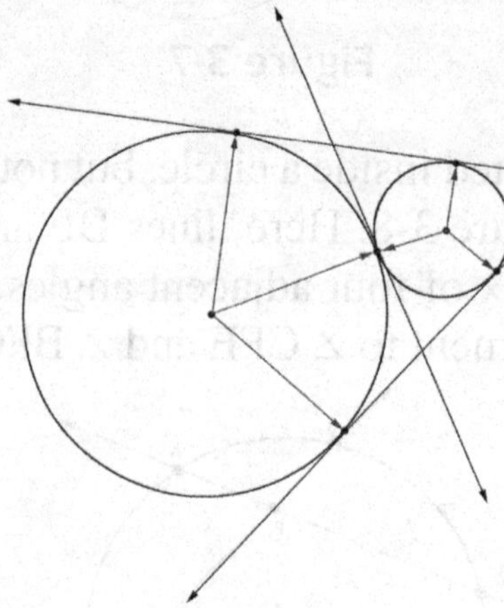

Figure 3-10

The circle as a conic section

A final word regarding circles that you will see in an algebra 2 or precalculus course when the topic of *conic sections* is studied: Equation (3-1) below is that of a circle.

$$x^2 + y^2 = r^2 \quad (3\text{-}1)$$

The constant ***r*** is the radius. Note that this equation is not a function (it fails the "vertical line test," remember?)

Refer to Figure 3-11. One can easily imagine a plane cutting a cone where the plane is parallel to that of the base of the cone. The resulting shape is a circle; thus the circle is one of the so-called conic sections.*

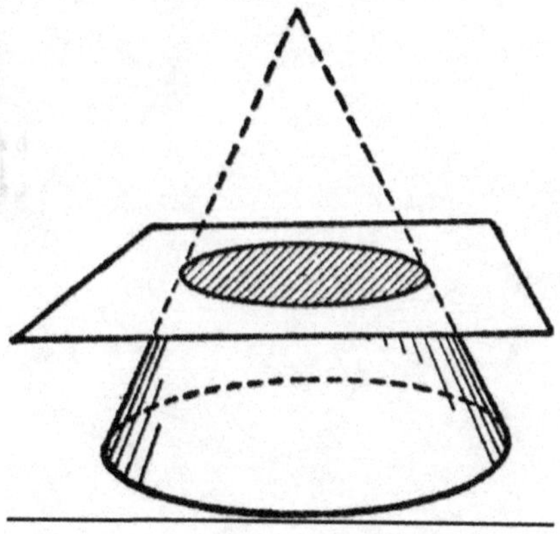

Figure 3-11

* Other conic sections include the ellipse, the parabola, and the hyperbola. Each results from the appropriate "tilting" of the cutting plane as already discussed in Section 1.

Inscribed angles and intercepted arcs

Figure 3-12 illustrates the theorem that states the measure of an inscribed angle is half the measure of the intercepted (minor) arc. Note that the measure of central angle AOC is exactly that of the intercepted arc. The inscribed angle that has one leg as tangent to a circle with the other forming a chord measures half that of the intercepted arc, as shown in Figure 3-13.

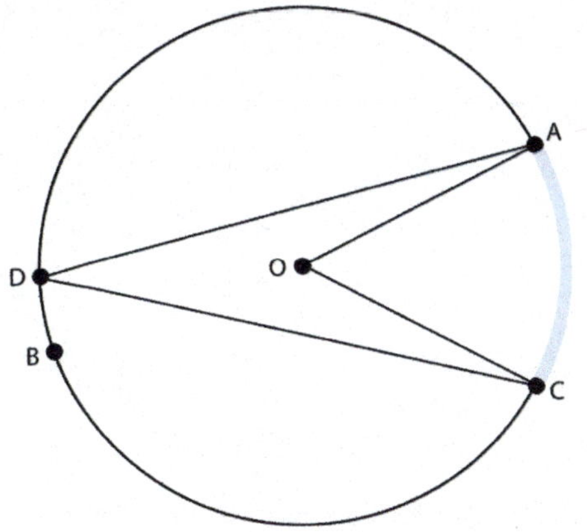

Figure 3-12

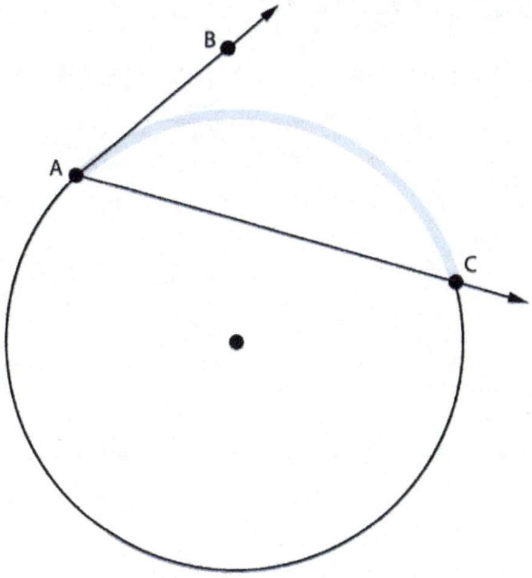

Figure 3-13

THE MATH TUTOR'S FRIEND
Pythagorean Theorem
similarity
congruence
It's easy !!
secant
transversal
Mr. Math
incenter
Postulates & Theorems

4. POSTULATES & THEOREMS
TABLE OF CONTENTS

page

4. POSTULATES AND THEOREMS

Postulates

AA Similarity	If two angles of one triangle are equal in measure to two angles of another triangle, then the two triangles are *similar*.
Angle Addition	For any angle, the measure of the whole is equal to the sum of the measures of its non-overlapping parts.
Arc Addition	The measure of an arc formed by two *adjacent arcs* is the is the sum of the measures of the two arcs.
Area Addition	The area of a region is equal to the sum of the areas of its non-overlapping parts
ASA Congruence	If two angles and the included side of one triangle are congruent to two angles and the included side of another triangle, then the triangles are congruent.
Corresponding Angles	If two parallel lines are intersected by a transversal, then the corresponding angles are equal in measure .
Intersecting-lines	If two lines intersect, then they intersect at exactly one point.
Parallel-lines	Through a point not on a given line, there is one and only one line parallel to the given line.
SAS Congruence	If two sides and the included angle of one triangle are equal in measure to the corresponding sides and angle of another triangle, then the triangles are congruent.
Segment Addition	For any segment, the measure of the whole is equal to the sum of the measures of its non-overlapping parts.
SSS Congruence	If three sides of one triangle are equal in measure to the corresponding sides of another triangle, then the triangles are congruent.
Two-point	Through any two points there is exactly one line.

Algebra properties

Addition Property of Equality	If the same number is added to equal numbers, then the sums are equal.
Subtraction Property of Equality	If the same number is subtracted from equal numbers, then the differences are equal.
Multiplication Property of Equality	If equal numbers are multiplied by the same number, then the products are equal.
Multiplication Property of Inequality	If both sides of an inequality ($<$ or $\leq$ or $>$ or $\geq$) are multiplied by the same positive number, then the inequality sense is unchanged. For inequalities multiplication by a negative number reverses the sense of the inequality.
Division Property of Equality	If equal numbers are divided by the same non-zero number, then the quotients are equal.
Division Property of Inequality	If both sides of an inequality ($<$ or $\leq$ or $>$ or $\geq$) are divided by the same positive number, then the sense of the inequality is unchanged. For inequalities division by a negative number reverses the sense of the inequality.
Reflexive Property of Equality	A number is equal to itself. $x = x$
Symmetric Property of Equality	If $a = b$ then $b = a$.
Asymmetric Property of Inequality	If $a < b$ then $b > a$.
Substitution Property	If $x = y$ then y may be substituted for x in any expression.
Transitive Property	If $a = b$ and $b = c$ then $a = c$. If $a > b$ and $b > c$ then $a > c$. ("$<$" may be replaced by "$\leq$" or "$>$" or "$\geq$")

Theorems

30°-60°-90° Triangle Theorem	In a 30°-60°-90° triangle, the hypotenuse length is 2 times the length of the shorter leg, and the length of the longer leg is the length of the shorter leg times the square root of 3.

45°-45°-90° Triangle Theorem	In a 45°-45°-90° triangle, both legs are congruent, and the length of the hypotenuse is the length of a side times the square root of 2.
AAS Congruence Theorem	If two angles and a non-included side of one triangle are equal in measure to the corresponding angles and side of another triangle, then the triangles are congruent.
Alternate Exterior Angles Theorem	If two parallel lines are intersected by a transversal, then alternate exterior angles are equal in measure.
Alternate Exterior Angles Converse	If two lines are intersected by a transversal and alternate exterior angles are equal in measure, then the lines are parallel.
Alternate Interior Angles Theorem	If two parallel lines are intersected by a transversal, then alternate interior angles are equal in measure.
Alternate Interior Angles Converse	If two lines are intersected by a transversal and alternate interior angles are equal in measure, then the lines are parallel.
Altitude to Hypotenuse Theorem	If the altitude is drawn to the hypotenuse of a right triangle, then the two triangles formed are similar to the original triangle and to each other.
Angle Bisector Theorem	If a point is on the bisector of an angle, then it is equidistant from the sides of the angle.
Angle Bisector Converse	If a point in the interior of an angle is equidistant from the sides of the angle, then it is on the bisector of the angle.
Angle-Opposite-Side Theorem	If two sides of a triangle are not congruent, then the larger angle is opposite the longer side.
Centroid Theorem	The centroid of a triangle is located 2/3 of the distance from each vertex to the midpoint of the opposite side.
Chord-Bisector Theorem	In a circle, if a radius (or diameter) is perpendicular to a chord, then it bisects the chord and its arc.
Chord-Bisector Converse	In a circle, the perpendicular bisector of a chord is a radius (or diameter).

Chord-Chord Product Theorem	If two chords intersect in the interior of a circle, then the products of the lengths of the segments of the chords are equal.
Circumcenter Theorem	The *circumcenter* of a triangle is equidistant from the vertices of the triangle.
Common Segments Theorem	Given collinear points A, B, C, D arranged as shown, if AB= CD then AC = BD. 
Congruent Central Angles Theorem	In a circle or congruent circles: congruent central angles have congruent chords, congruent chords have congruent arcs and congruent arcs have congruent central angles.
Congruent Complements Theorem	If two angles are complements of the same angle, then they are congruent.
Congruent Right Angles Theorem	If two congruent angles are supplementary, then each is a right angle.
Congruent Supplements Theorem	If two angles are supplements of the same angle, then they are congruent.
Congruent Tangents Theorem	If two segments are tangent to a circle from the same external point then they are congruent.
Corresponding Angles Converse	If two lines are intersected by a transversal and corresponding angles are equal in measure, then the lines are parallel.
CPCFC	Congruent parts of congruent figures are congruent.
CPCTC	Congruent parts of congruent triangles are congruent.
Diagonal Angle Bisection Theorem	If one diagonal of a parallelogram bisects a pair of opposite angles, then the parallelogram is a rhombus.
Equilateral Triangle Theorem	If a triangle is equilateral, then it is equiangular.
Equilateral Triangle Converse	If a triangle is equiangular, then it is equilateral.
Equilateral Triangle Corollary	Each angle of an equiangular triangle is 60°.

Exterior Angle Theorem	An exterior angle of a triangle is equal in measure to the sum of the measures of its two remote interior angles.
Geometric Means Corollary A	The length of the altitude to the hypotenuse of a right triangle is the geometric mean of the lengths of the two segments of the hypotenuse.
Geometric Means Corollary B	The length of a leg of a right triangle is the geometric mean of the lengths of the hypotenuse and the segment of the hypotenuse adjacent to that leg.
Hinge Theorem	If two sides of one triangle are congruent to two sides of another triangle and the third sides are not congruent, then the longer third side is across from the larger included angle.
Hinge Theorem Converse	If two sides of one triangle are congruent to two sides of another triangle and the third sides are not congruent, then the larger included angle is opposite from the longer third side.
HL Congruence Theorem	If the hypotenuse and a leg of a right triangle are congruent to the hypotenuse and a leg of another right triangle, then the triangles are congruent.
Incenter Theorem	The incenter of a triangle is equidistant from the sides of the triangle.
Inscribed Angle Theorem	The measure of an inscribed angle is half the measure of its intercepted arc.
Inscribed Angle Corollary	If inscribed angles of a circle intercept the same arc or are subtended by the same chord or arc, then the angles are congruent.
Inscribed Quadrilateral in Circle Theorem	If a quadrilateral is inscribed in a circle, then its opposite angles are supplementary.
Isosceles Trapezoid Base Angles Theorem	If a quadrilateral is an isosceles trapezoid, then each pair of base angles are congruent.
Isosceles Trapezoid Base Angles Converse	If a trapezoid has one pair of congruent base angles, then the trapezoid is isosceles.

Isosceles Trapezoid Diagonals Theorem	A trapezoid is isosceles if and only if its diagonals are congruent.
Isosceles Triangle Theorem	If two sides of a triangle are equal in measure, then the angles opposite those sides are equal in measure.
Isosceles Triangle Converse	If two angles of a triangle are equal in measure, then the sides opposite those angles are equal in measure.
Kite Diagonals Theorem	If a quadrilateral is a kite then its diagonals are perpendicular.
Kite Opposite Angles Theorem	If a quadrilateral is a kite then exactly one pair of opposite angles are congruent.
Linear Pair Theorem	If two angles form a linear pair, then they are supplementary.
Parallel Lines Theorem	In a coordinate plane, two non-vertical lines are parallel IFF they have the same slope.
Parallelogram Consecutive Sides Theorem	If one pair of consecutive sides of a parallelogram are congruent, then the parallelogram is a rhombus.
Parallelogram Congruent Diagonals Theorem	If the diagonals of a parallelogram are congruent, then the parallelogram is a rectangle.
Parallelogram Consecutive Angles Theorem	If a quadrilateral is a parallelogram, then its consecutive angles are supplementary.
Parallelogram Diagonals Theorem	If a quadrilateral is a parallelogram, then its diagonals bisect each other.
Parallelogram Diagonals Converse	If the diagonals of a quadrilateral bisect each other, then the quadrilateral is a parallelogram.
Parallelogram Opposite-Angles Theorem	If a quadrilateral is a parallelogram, then its opposite angles are congruent.
Parallelogram Opposite-Sides Theorem	If a quadrilateral is a parallelogram, then its opposite sides are congruent.
Perpendicular Lines Theorem	If two intersecting lines form a linear pair of congruent angles, then the lines are perpendicular.
Perpendicular Lines Theorem	In a coordinate plane, two non-vertical lines are perpendicular IFF the product of their slopes is -1.

Perpendicular Lines to Transversal Theorem	If two lines are perpendicular to the same transversal, then they are parallel.
Perpendicular Segment Bisector Theorem	If a point is on the perpendicular bisector of a segment, then it is equidistant from the end-points of the segment.
Perpendicular Segment Bisector Converse	If a point is the same distance from both the endpoints of a segment, then it lies on the perpendicular bisector of the segment.
Perpendicular Transversal Theorem	If a transversal is perpendicular to one of two parallel lines, then it is perpendicular to the other one.
Polygon Angle Sum Theorem	The sum of the interior angle measures of a convex polygon with *n* sides is given by $(n - 2) \times 180°$.
Polygon Exterior Angle Sum Theorem	The sum of the exterior angle measures, one angle at each vertex, of a convex polygon is 360°.
Proportional Perimeters, Areas Theorem	If the similarity ratio of two similar figures is a:b then the ratio of their perimeter is a:b; the ratio of their areas is a^2:b ; the ratio of their volumes is a^3:b^3.
Pythagorean Inequalities Theorem	In triangle ABC, c is the length of the longest side. If $c^2 > a^2 + b^2$, then ABC is an *obtuse triangle.* If $c^2 < a^2 + b^2$, then ABC is acute.
Pythagorean Theorem	In any right triangle, the square of the length of the hypotenuse is equal to the sum of the square of the lengths of the legs.
Pythagorean Converse	If the sum of the squares of the lengths of two sides of a triangle is equal to the square of the length of the third side, then the triangle is a *right triangle*.
Quadrilateral-Parallelogram Theorem	If one pair of opposite sides of a quadrilateral are parallel and congruent, then the quadrilateral is a *parallelogram*.
Rectangle Diagonals Theorem	If a parallelogram is a rectangle, then its diagonals are congruent.

Rhombus Diagonals Theorem	If a parallelogram is a rhombus, then each diagonal bisects a pair of opposite angles.
Rhombus Diagonals Converse	If the diagonals of a parallelogram are perpendicular, then the parallelogram is a *rhombus*.
Right Angle Congruence Theorem	All right angles are congruent.
Right-Angle Parallelogram Theorem	If one angle of a parallelogram is a right angle, then the parallelogram is a rectangle.
Same-side Interior Angles Theorem	If two parallel lines are intersected by a *transversal*, then same-side interior angles are supplementary.
Same-side Interior Angles Converse	If two lines are intersected by a transversal and same-side interior angles are supplementary, then the lines are parallel.
SAS Similarity Theorem	If two sides of one triangle are proportional to two sides of another triangle and their included angles are congruent, then the triangles are similar.
Secant-Chord Interior Intersection Theorem	If two secants or chords intersect in the interior of a circle, then the measure of each angle formed is half the sum of the measures of the intercepted arcs.
Secant-Chord Tangent Intersection Theorem	If a tangent and a secant (or chord) intersect on a circle at the point of tangency, then the measure of the angle formed is half that of its intercepted arc.
Secant-Secant Product Theorem	If two secants intersect in the exterior of a circle, then the product of the lengths of one secant segment and its external segment equals the product of the lengths of the other secant segment and its external segment.
Secant-Tangent External Intersection Theorem	If a tangent and a secant, two tangents or two secants intersect in the exterior of a circle, then the measure of the angle formed is half the difference of the measure of its intercepted arc.

Secant-Tangent Product Theorem	If a secant and a tangent intersect in the exterior of a circle, then the product of the lengths of the secant segment and its external segment equals the length of the tangent segment squared.
Semicircle Inscribed Angle Theorem	An inscribed angle subtends a semicircle IFF the angle is a right angle.
Side Opposite-Angle Theorem	If two angles of a triangle are not congruent, then the longer side is opposite the larger angle.
SSS Similarity Theorem	If the three sides of one triangle are proportional to the three corresponding sides of another triangle, then the triangles are similar.
Tangent-Radius Theorem	If a line is tangent to a circle, then it is perpendicular to the radius drawn to the point of tangency.
Tangent-Radius Converse	If a line is perpendicular to a radius of a circle at a point on the circle, then the line is tangent to the circle.
Third Angles Theorem	If two angles of one triangle are congruent to two angles of another triangle, then the third pair of angles are congruent.
Trapezoid Midsegment Theorem	The *midsegment of a trapezoid* is parallel to each base, and its length is one half the sum of the lengths of the bases.
Triangle Angle Bisector Theorem	An *angle bisector of a triangle* divides the opposite sides into two segments whose lengths are proportional to the lengths of the other two sides.
Triangle Inequality Theorem	The sum of any two side lengths of a triangle is greater than the third side length.
Triangle Midsegment Theorem	A *midsegment of a triangle* is parallel to a side of triangle, and its length is half the length of that side.

Triangle Proportionality Theorem	If a line parallel to a side of a triangle intersects the other two sides, then it divides those sides proportionally.
Triangle Sum Theorem	The sum of the measure of the angles of a triangle is 180°.
Triangle Sum Corollary	The acute angles of a right triangle are complementary.
Two-Transversal Proportionality Theorem	If three or more parallel lines intersect two transversals, then they divide the transversals proportionally.
Vertical Angles Theorem	Vertical angles are congruent (equal in measure).

THE MATH TUTOR'S FRIEND
DISJUNCTION
TAUTOLOGY
p ~p
T F
F T
It's easy !!
NEGATION
IMPLICATION
CONTRA-POSITIVE
Mr. Math
LOGIC &
TRUTH TABLES

5. LOGIC & TRUTH TABLES
TABLE OF CONTENTS

page

•

5. LOGIC AND TRUTH TABLES

Introduction

This section discusses *two-valued logic*, called the *Law of the Excluded Middle*, that is, the only two states are "True" or "False" (T or F) - there is nothing in between, hence the excluded middle.

Logical Statements

Logic deals with *statements* - sentences that are either true or false. For example, "the cloud is white" is a statement because it is either true or false, whereas "how are you?" is not a (logical) statement. Let a *simple statement* (also called *arbitrary statement*) be represented by the symbol p; * thus p has truth value "T" (true) or "F" (false). Now if p is true, then its *negation* (symbolized by $\sim p$) is false. Let's summarize this with a *truth table*:**

p	$\sim p$
T	*F*
F	T

A negation of a statement is simply the *opposite truth value* of that statement.

Symbols used in logic

Some of the more common symbols used in logic are " $\sim$ " (or " $\neg$ ") for negation, and the so-called *logical connectives*: " $\rightarrow$ " (or " $\Rightarrow$ ") for *conditional* or *implication*, (" $\leftrightarrow$ ") (or " $\Leftrightarrow$ " or " $\equiv$ ") for *bi-conditional* or *equivalence*, (" $\wedge$ ") for *conjunction*, and (" $\vee$ ") for *disjunction*. These symbols are used to form *compound statements*.

Compound statements

Just as statements are either T or *F*, so are compound statements themselves T or *F* depending upon the truth or falsity of the included simple statements. These are summarized in the table on the next page.

* The symbol p is a variable, such as x in algebra. A statement may be symbolized by other letters such as c, q, r, *etc...*

** We italicize "*F*" in the truth tables to make it more distinguishable from "T."

Symbol	Example	Name
$\wedge$	$p \wedge q$ "p AND q"	conjunction
$\vee$	$p \vee q$ "p OR q"	disjunction
$\rightarrow$	$p \rightarrow q$ "if p then q"	conditional
Symbol	**Example**	**Name**
$\Rightarrow$	$p \wedge q \Rightarrow p \vee q$ "$p \wedge q$ IMPLIES $p \vee q$"	conditional between compound statements
$\leftrightarrow$	$p \leftrightarrow q$ "p if and only if q"	biconditional
$\Leftrightarrow$ or $\equiv$	"p is EQUIVALENT to q" $(p \rightarrow q) \Leftrightarrow (\sim p \rightarrow \sim q)$ [contraposition]	equivalent statements

There are many symbols employed in logic documents. For example, the symbol "$\vee$" is an *inclusive* ***OR***; that is, $p \vee q$ means "either p or q or both." There is also an *exclusive* ***OR***, one symbol for which is "$\veebar$." That is, $p \veebar q$ means "either p or q but not both."

Consider the disjunction $p \vee q$ and look at its truth table:

p	q	$p \vee q$
T	T	T
T	F	T
F	T	T
F	F	F

Note that the table displays four cases. If there are three statements, say p, q, and r, then there will be eight cases. In general, if there are n statements then there will be 2^n cases to list all possible combinations of T and F.

Addition Law $p \Rightarrow p \vee q$

Associative Law $p \vee (q \vee r) \Leftrightarrow (p \vee q) \vee r$

$p \wedge (q \wedge r) \Leftrightarrow (p \wedge q) \wedge r$

Commutative Law $p \wedge q \Leftrightarrow q \wedge p$

$p \vee q \Leftrightarrow q \vee p$

Rules of inference (cont'd)

Constructive dilemma	$(p \rightarrow q) \wedge (r \rightarrow s) \Rightarrow (p \vee r) \rightarrow (q \vee s)$
	$\Rightarrow (p \wedge r) \rightarrow (q \wedge s)$
Contraposition	$(p \rightarrow q) \Leftrightarrow (\sim p \rightarrow \sim q)$
De Morgan's Law	$\sim(p \wedge q) \Leftrightarrow (\sim p \vee \sim q)$
	$\sim(p \vee q) \Leftrightarrow (\sim p \wedge \sim q)$
Destructive dilemma	$(p \rightarrow q) \wedge (r \rightarrow s) \Rightarrow (\sim q \vee \sim s) \rightarrow (\sim p \vee \sim r)$
	$\Rightarrow (\sim p \wedge \sim s) \rightarrow (\sim p \wedge \sim s)$
Disjunctive Syllogism	$(p \vee q) \wedge \sim p \Rightarrow q$
Distributive Law	$p \wedge (q \vee r) \Leftrightarrow (p \wedge q) \vee (p \wedge r)$
	$p \vee (q \wedge r) \Leftrightarrow (p \vee q) \wedge (p \vee r)$
Double Negative	$\sim(\sim p) \Leftrightarrow p$
Modus Ponens	$(p \rightarrow q) \wedge p \Rightarrow q$
Modus Tollens	$(p \rightarrow q) \wedge \sim q \Rightarrow \sim p$
Reductio ad Absurdum	$(p \rightarrow q) \Leftrightarrow (p \wedge \sim q) \rightarrow (q \wedge \sim q)$
Simplification	$(p \wedge q) \Rightarrow p$
	$(p \wedge q) \Rightarrow q$
Transitivity	$(p \rightarrow q) \wedge (q \rightarrow r) \Leftrightarrow (p \rightarrow r)$

Implication

The logical statement $(p \rightarrow q)$ is an implication; i.e., "p implies q." Implication may also be indicated by the logical statement $(p \wedge q) \Rightarrow p$. We shall use the symbol "$\rightarrow$" between simple statements and the symbol "$\Rightarrow$" between compound statements and between a simple statement and a compound statement. A statement of implication is also called a *conditional statement*. Before addressing the conditional statement, let's look at implication:

Statement $p \rightarrow q$. Switch the variable and we have the *Converse* $q \rightarrow p$
Negate the components of the statement to get the *Inverse* $\sim p \rightarrow \sim q$.
Switch the negated components to get the *Contrapositive* $\sim q \rightarrow \sim p$
For example, say p is the simple statement "you are over 18" and q is "you can vote." The negations of these, of course, are "you are not over 18" and "you cannot vote," respectively. Now consider the implication $p \rightarrow q$ which is "If you are over 18 then you can vote." Its converse is "if you can vote then you are over 18. The inverse is "if you are not over 18 then you cannot vote," and the contrapositive is "If you cannot vote then you are not over 18." These compound statements do make sense; but this is not always true; for example, let p be the simple statement "the car is green" and q be the simple statement "it is raining in New York," the converse, inverse, and contrapositive statements are all logically true but make no real sense.

The conditional statement

Let p, q, and r be simple statements. The compound statement $p \Rightarrow (q \wedge \sim r)$ is an implication and has the following truth values:

p	q	r	$\sim r$	$q \wedge \sim r$	$p \Rightarrow (q \wedge \sim r)$
T	T	T	F	F	F
T	T	F	T	T	T
T	F	T	F	F	F
T	F	F	T	F	F
F	T	T	F	F	T
F	T	F	T	T	T
F	F	T	F	F	T
F	F	F	T	F	T

As shown above, a "truth-table" analysis can get tedious. An easier way to deal with this is to apply the truth or falsity symbols (T or F) to the ***connectives*** as shown below. The first row shows the order of applying the steps of the *alternate truth-table analysis* for the contrapositive $(p \rightarrow q) \Leftrightarrow (\sim p \rightarrow \sim q)$.

1	3	1	4	2	3	2
p	$\rightarrow$	q	$\Leftrightarrow$	$\sim q$	$\rightarrow$	$\sim p$
T	T	T	T	F	T	F
T	F	F	T	T	F	F
F	T	T	T	T	T	T
F	T	F	T	T	T	T

Analyzing the table we first insert all possible combinations of T and *F* for p and q. Next, in the columns labeled "2," we negate all the truth values of columns labeled "1." Step 3 is to determine the truth values of the implications in the columns labeled "3." Finally, in Step 4 we look at the implications for the columns labeled "3," and see that the truth values are identical in both of these columns, so insert "T" for which proves the equivalence ("⇔"). The contrapositive is a *tautology*. The alternate truth-table analysis is useful for long compound statements.

Tautologies and contradictions

Theorem: Let t be a *tautology*, c a *contradiction*, and p an *arbitrary statement*, then,

(a) $p \wedge t \Leftrightarrow p$ (b) $p \vee c \Leftrightarrow p$ (c) $c \Rightarrow p$, and $p \Rightarrow t$.

$p \vee t \Leftrightarrow t$ $p \wedge c \Leftrightarrow c$

You can prove these easily using truth tables. Here is a table of tautologies:

TAUTOLOGIES

1	$[(p \wedge q) \wedge r] \Leftrightarrow [p \wedge (q \wedge r)]$	Associative Law for ∧
2	$[(p \vee q) \vee r] \Leftrightarrow [p \vee (q \vee r)]$	Associative Law for ∨
3	$(p \leftrightarrow q) \Leftrightarrow [(p \rightarrow q) \wedge (q \rightarrow \mathrm{p})]$	Biconditional definition
4	$(p \wedge q) \Leftrightarrow (q \wedge p)$	Commutative Law for ∧
5	$(p \vee q) \Leftrightarrow (q \vee p)$	Commutative Law for ∨
6	$(p \leftrightarrow q) \Leftrightarrow (\sim p \vee q)$	Conditional disjunction
7	$(p \wedge q) \rightarrow p$ $(p \wedge q) \rightarrow q$	conjunction decomposition
8	$\sim(p \wedge \sim p)$	contradiction
9	$(p \rightarrow q) \Leftrightarrow (\sim q \rightarrow \sim p)$	contrapositive
10	$\sim(p \vee q) \Leftrightarrow (\sim p \wedge \sim q)$ $\sim(p \wedge q) \Leftrightarrow (\sim p \vee \sim q)$	DeMorgan's law
11	$p \rightarrow (p \vee q)$ $q \rightarrow (p \vee q)$	disjunction construction
12	$[(p \vee q) \wedge \sim p] \Rightarrow q$	disjunctive syllogism
13	$[p \wedge (q \vee r)] \Leftrightarrow [(p \wedge q) \vee (p \wedge r)]$ $[p \vee (q \wedge r)] \Leftrightarrow [(p \vee q) \wedge (p \vee r)]$	Distributive law
14	$\sim\sim p \Leftrightarrow p$	double negation
15	$[(p \rightarrow q) \wedge (q \rightarrow r)] \Rightarrow (p \rightarrow r)$	Law of syllogism
16	$p \vee \sim p$	Law of the excluded middle
17	$[(p \rightarrow q) \wedge p] \Rightarrow q$	*modus ponens* (forward reasoning)
18	$[(p \rightarrow q) \wedge \sim q] \Rightarrow \sim p$	*modus tollens*
19	$(p \vee p) \Leftrightarrow p$ $(p \wedge p) \Leftrightarrow p$	Simplification

Fallacies

Logical analysis does not tell us whether a hypothesis is true; it addresses the validity of an argument, not the truth or falsity of a proposition. Logic merely helps determine if the conclusion follows from the hypothesis. There are many fallacies and types of fallacies, both formal and informal, way beyond the scope of this book. Here are a few:

Formal fallacies

Assuming the consequent

"Successful authors are writers; I am a writer, therefore I am a successful author." Symbolically, this is expressed as $(p \rightarrow q) \wedge q \Rightarrow p$ which can be easily disproved with a truth table.

Informal fallacies

Paradox

" 'Can't' is not a word in my vocabulary." If the speaker uses the word then it must be in the vocabulary.

"I am a liar." If the statement is true then the speaker cannot be a liar.

Overprecision

Here's an example of the informal fallacy of overprecision: "I visited a museum last year and saw the fossil skeleton of a dinosaur that the sign stated was 3 million years old, so today, the bones are 3,000,001 years old."

•

SYMBOLS USED IN LOGIC AND MATH

SYMB	NAME/MEANING	EXAMPLE
$\vee$	Disjunction	$p \vee q$ *p* or *q* (or both)
$\veebar$	Exclusive Disjunction	$p \veebar q$ *p* or *q* (but not both)
$\wedge$	Conjunction	$p \wedge q$ *p* and *q*
~	Negation	$\sim p$ the statement *p* is False
$\rightarrow$	Conditional	$p \rightarrow q$ if *p* then *q*
$\Rightarrow$	Conditional (*e.g.*, of complex statements)	$p \wedge q \Rightarrow q$
$\leftrightarrow$	Biconditional; also abbreviated as "iff"	$p \leftrightarrow q$ *p* if and only if *q*
$\Leftrightarrow$ or $\equiv$	Equivalence (*e.g.*, of complex statements)	$p \wedge q \Leftrightarrow q \wedge p$
	QUANTIFIERS	
$\in$	"is an element (or member) of	"$2 \in \{2,4,6,8,\ldots\}$
$\ni$	"such that"	quantifier
$\exists$	"there exists"	quantifier
$\forall$	"for all"	$f(x) = g(x)\ \forall x$
$\therefore$	"Therefore"	
Q.E.D.	*Quod Erat Demonstrandum* "the proof is complete"	

INDEX

Symbols

A

B

C

• • •

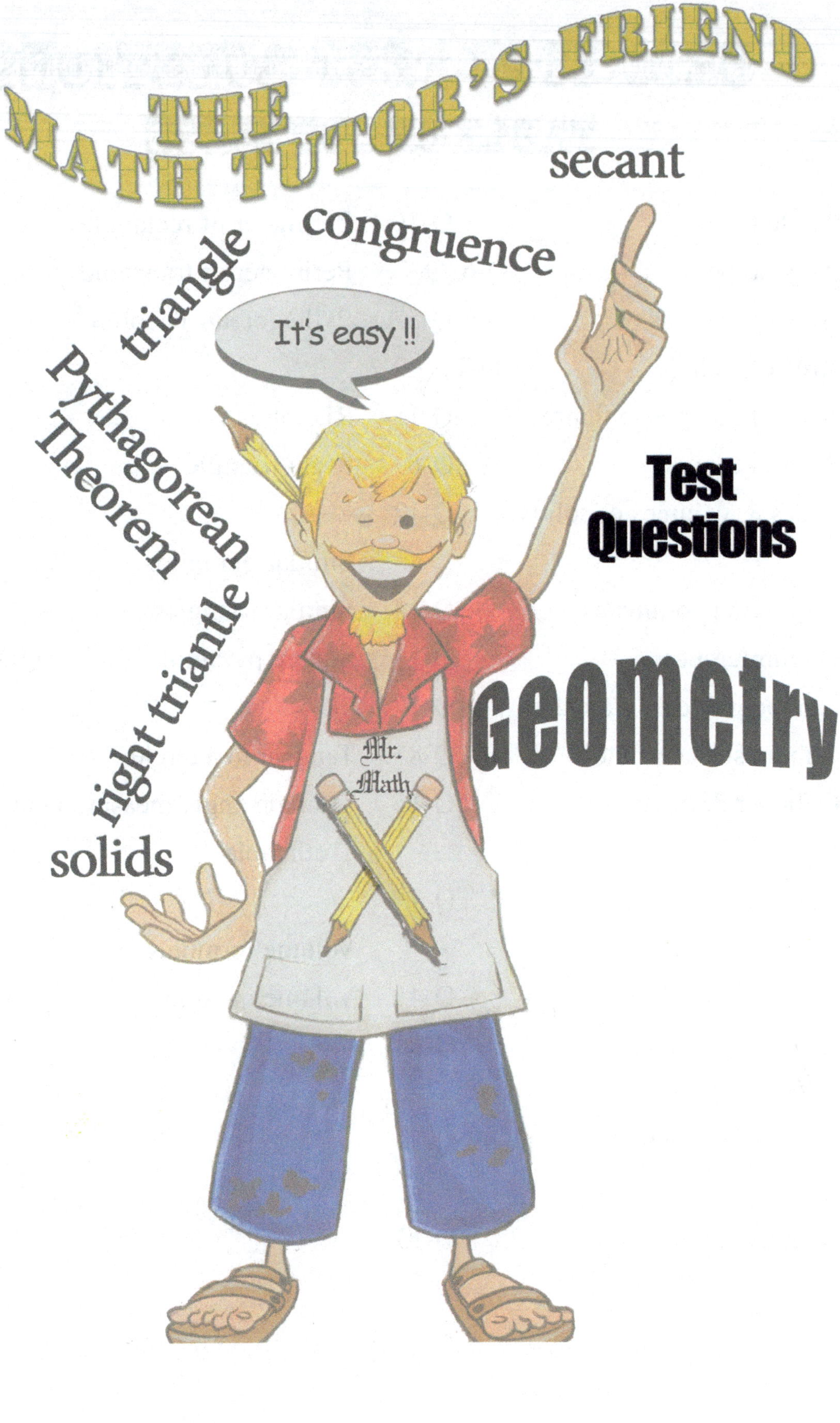
THE MATH TUTOR'S FRIEND
secant
congruence
triangle
It's easy !!
Pythagorean Theorem
Test Questions
right triantle
Geometry
Mr. Math
solids

GEOMETRY TEST QUESTIONS
TABLE OF CONTENTS

• How many square inches of tin metal are needed to make a single soup can that is five inches high and three inches in diameter?

A. 15π

B. 4.50π

C. 16π

D. 19.50π

E. None of the above

• • Answer D.

Total area is $A_{TOP} + A_{BOTTOM} + A_{SIDE}$. The radius is $r = d/2 = 1.5$ inches.

$A_{TOP} = A_{BOTTOM} = \pi r^2 = \pi \times (1.5)^2 = 2.25\pi$ square inches

$A_{SIDE} = 5 \times \text{Circumference} = 5 \times \pi d = 5 \times 3 \times \pi = 15\pi$ square inches.

The total area of the can is $2 \times 2.25\pi + 15\pi = 4.50\pi + 15\pi = 19.50\pi$ square inches (an exact answer).

• ♥ •

• The radius of a circle is 4 inches. What is the circle's circumference?

A. Not enough data to solve the problem

B. 2π inches

C. 4π inches^2 (square inches)

D. 8π inches

E. None of the above

• • Answer D.

Radius = r. Circumference = $C = \pi d = 2\pi r = 2\pi \times 4 = 8\pi$ inches.

• A circle has an area of 36π square centimeters. What is the length of the circle's diameter?

A. 6 centimeters

B. 6π centimeters

C. 12 centimeters

D. 36 centimeters

E. None of the above

• • Answer C. Area = A = $36\pi = \pi r^2$

$r^2 = 36$ $r = 6$ Diameter = $2r = 2 \times 6 = 12$ centimeters

• ♥ •

• The ratio of apothems of a square pyramid is 4.5 times that of another similar pyramid. If the volume of the first pyramid is 5,400 cubic meters, what is the approximate volume of the second pyramid?

A. 729 m^3

B. 30 m^3

C. 1200 m^3

D. 60 m^3

E. None of the above

• • Answer D. The ratio, a:b is 4.5:1 or 9:2. The ratio of volumes is $a^3 : b^3$. Here, $a^3 = 5{,}400$ m^3. We must solve for b^3 : $9^3/2^3 = 729/8 = 5400/x$ from which $x \approx 59.3$ m^3

• If each corner bend requires 1/4" of wire length, how much excess wire is there after bending a 45-inch aluminum wire into the shape of a rectangle whose length is twice its width of 7 cm?

A. 0 inches (no excess wire)

B. 2 inches

C. 3 inches

D. 4 inches

E. The wire is too short to do this

• • Answer B.

The rectangle has two 7-inch sides and two 14-inch sides (2×7).

The perimeter is 2×7 + 2×14 = 42 inches plus 4 ×1/4" = 43 inches, leaving two excess inches.

• ♥ •

• A cube has a surface area of 96 square inches. The volume of the cube is ___ in^3 (cubic inches).

A. 96

B. 46

C. 64

D. 16

E. None of the above

• • Answer C.

Let x be the length of a side of the cube. There are six faces, each of area x^2, so the total surface area is $6x^2 = 96$ from which $x^2 = 16$ and $x = 4$. The volume of this cube is $x^3 = 4^3 = 64$ cubic inches.

• If the height of a triangle is five inches less than the length of its base, and if the area of the triangle is 52 square inches, find the height, *h*, of the triangle if the base, b, is 13 inches.

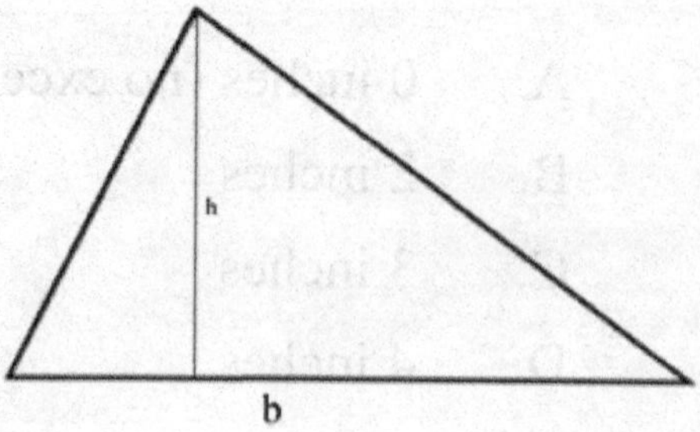

A. 10 inches

B. 12 inches

C. 13 inches

D. 8 inches

E. None of the above

• •Answer D.

Area, $A = 52 = ½\, bh = ½ \times 13h$. Solving for *h*: $h = (2\times52)/13 = 8$ inches.

or, more directly, $h = b - 5 = 13 - 5 = 8$ inches.

• ♥ •

• One of the sides of a right triangle is 6 inches and the hypotenuse is 10 inches, find the length of the remaining side.

A. 8 inches

B. 10 inches

C. 11 inches

D. 12 inches

E. None of the above

• • Answer A.

This is a right triangle so we can use the Pythagorean Theorem: $c^2 = a^2 + b^2$

$102 = 62 + b^2$ or $100 = 36 + b^2$ from which $b^2 = 100 - 36 = 64$ and so $b = 8$ inches. This is a similar triangle to a 3-4-5 triangle (scale factor = 2).

• A circular swimming pool with a diameter of 24 feet has a 4-foot uniform width deck built around it. What is the area of the deck?

A. 16π square feet
B. 112π square feet
C. 756π square feet
D. 576π square feet
E. None of the above

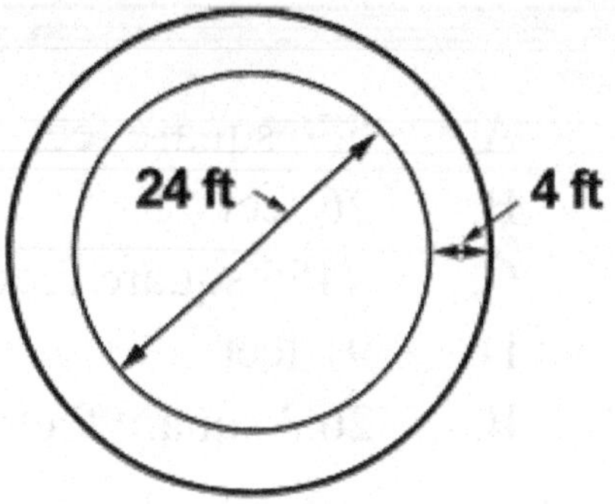

• • Answer B.

The pool radius is 12 ft. The radius of the pool & deck is 16 ft.

$$A_{pool} = \pi \times 122 = 144\pi \text{ square feet}$$

$$A_{pool} + A_{deck} = \pi \times 162 = 256\pi \text{ square feet}$$

Therefore, the area of the deck is $(256 - 144)\pi = 112\pi$ square feet.

• ♥ •

• The largest angle of a triangle is twice the size of one other angle and ⁴⁄₃ the size of the remaining angle. The three angle measurements are

A. 30°, 60°, 90°
B. 40°, 60°, 80°
C. 60°, 60°, 60°
D. 50°, 60°, 70°
E. None of the above

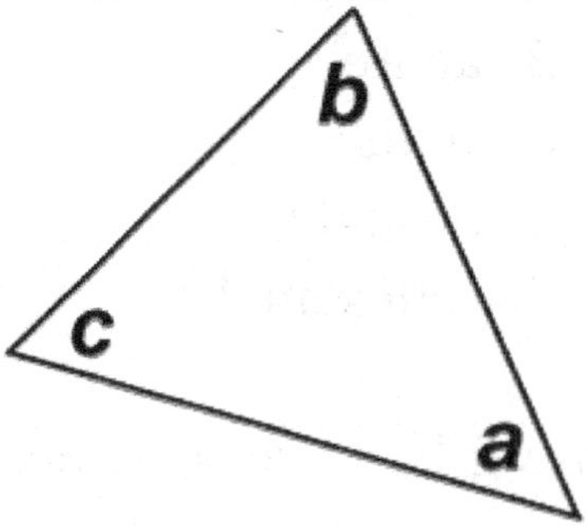

• • Answer B.

Let $a = m\angle a$, $b = m\angle b$, and $c = m\angle c$,
then $a + b + c = 180°$
We are given $b = 2a$ and $b = 4c/3$. Put everything in terms of a.
$4c/3 = 2a$ from which $c = 6a/4 = 3a/2$.
So, $a + 2a + 3a/2 = 360°$
Solve for a: $a = 40°$, $b = 80°$ and $c = 60°$.

• What is the area of the triangle shown below? Side lengths are in feet. Give an exact answer.

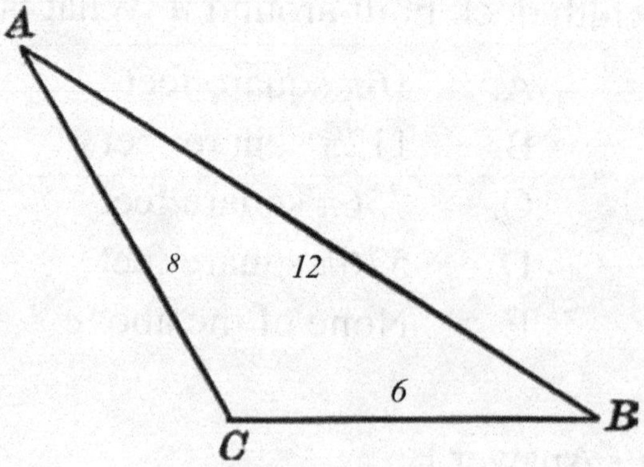

A. 576 square feet
B. 26 feet
C. $\sqrt{455}$ square feet
D. 91 feet
E. 26.3 square feet

• • Answer C. Use Herron's formula

s = (12 + 8 + 6)/2 = 26/2 = 13 feet

$$s = \frac{a+b+c}{2}$$

$$A = \sqrt{13(13-12)(13-8)(13-6)}$$

$$= \sqrt{13 \times 1 \times 5 \times 7} = \sqrt{455}$$

$$A = \sqrt{s(s-a)(s-b)(s-c)}$$

Note that Answer E is an approximate answer.

• ♥ •

• A 30° angle is a(n) ______ angle

A. obtuse
B. acute
C. right
D. straight
E. reflex angle

• • Answer B. It is an acute angle because its measurement is between 0° and 90°.

Definition:

(If $0° < m\angle\alpha < 90°$ then α is an acute angle.)
(If $m\angle\alpha = 90°$ then α is a right angle.)
(If $90° < m\angle\alpha < 180°$ then α is an obtuse angle.)
(If $m\angle\alpha = 180°$ then α is a straight angle.)
(If $m\angle\alpha > 180°$ then α is a reflex angle.)

• An angle of 90° is a(n) ______ angle

A. obtuse
B. acute
C. right
D. straight
E. oblique

• • Answer C.

A right angle is one of 90°. That is, if $m\angle A = 90°$, then $\angle A$ is a right angle.

• ♥ •

• A trapezoid has base lengths of 2 feet and 5 feet; the other two sides have lengths of 6 feet and 10 feet, respectively. The height, *h*, is 4 feet. The perimeter of this trapezoid is

A. 23 feet
B. 27 feet
C. 27 square feet
D. 10 feet
E. None of the above

• • Answer A. Height is not included in a perimeter measurement.

• Match the volume formula to name of the shape (more than one formula may apply to a given shape):

A. $4/3\ \pi r^3$	___	oblique cylinder
B. 2LW + 2LD + 2WD	___	box
C. πr^2	___	sphere
D. $\pi r^2 h$	___	circle

• • Answer A (sphere); B (box); C (circle) D (oblique cylinder)

• ♥ •

• Sean intends to wrap his mother's gift – a box measuring 5" × 20" × 15" He estimates that 10% additional paper is needed to allow for overlap. There is one 6 square-foot sheet of wrapping paper available. Is there enough paper to do the job?

- If so, how much excess paper will there be?
- If not, what size sheet (to the nearest square foot) will he need?

A. Yes; there is an excess of 12 square inches

B. Yes; there is an excess of 1.3 square feet

C. No; he will need a sheet of 8 square feet

D. No; he will need a sheet of 7 square feet

E. None of the above

• • Answer C. The area of the top and bottom is $2 \times 20 \times 15 = 600$ in ^{2}.
The area of both ends is $2 \times 5 \times 15 = 150$ in ^{2}.
The area of both sides is $2 \times 5 \times 20 = 200$ in ^{2}.
Total area is $600 + 150 + 200 = 950$ in ^{2}.
The 10% allowance is another 95 in^2 for a total of 1,045 in ^{2}, or about 7.3 ft^2 (1,045/144). Sean will need a sheet of wrapping paper that is 8 square feet (rounding up from 7.3).

• A rhombus with side length 16 inches is similar to a rhombus with 2 cm sides.

TRUE

FALSE

• • Answer: TRUE. All rhombi are similar

• ♥ •

• All of the angles formed by the intersection of the diagonals of a rhombus

A. are equal; each angle is 60°

B. may not be equal

C. are equal in pairs, two if which are 60°

D. are equal; each angle is 90°

E. None of the above

• • Answer D. The diagonals of a square or of a rhombus intersect at right angles.

• All edges of the square pyramid shown have length b. Which of the following best describes the faces of this solid figure?

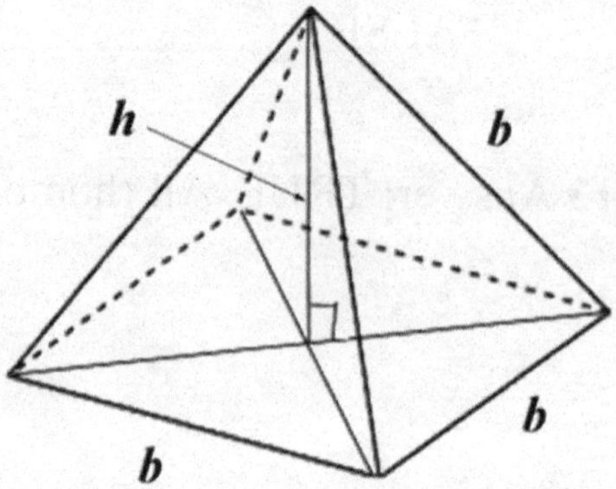

A. 4 equilateral triangles and a square

B. 8 triangles and a sphere

C. 8 triangles and a square

D. 2 triangles

E. None of the above

• • Answer A.

Each of the 4 triangular faces is an equilateral triangle with side-length b. The base is a square with side length b.

• ♥ •

• An angle of 130° is a(n) ______ angle

A. obtuse

B. acute

C. right

D. straight

E. oblique

• • Answer A. It is between 90° and 180°.

If $90° < m\angle A < 180°$, then $\angle A$ is an obtuse angle.

• An air-conditioning contractor must determine the volume of a home. A partition is to be built to allow only half of the attic to be air-conditioned. What is the volume needed? Round your answer to the next higher thousand cubic feet.

A. 1,000 cubic feet

B. 2,000 cubic feet

C. 3,000 cubic feet

D. 4,000 cubic feet

E. None of the above

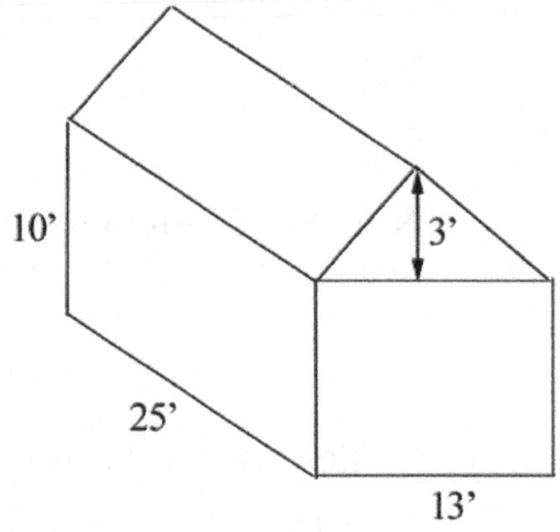

• • Answer D.

Attic volume is $\frac{1}{2} \times 3 \times 13 \times 25 = 487.5$ ft^3. Half of that is 243.75 ft^3.

The volume of the "box" is $10 \times 25 \times 13 = 3{,}250$ ft^3.

The total volume is $3{,}250 + 243.75 = 3{,}493.75$ ft^3 which rounds up to 4,000 ft^3.

• ♥ •

• The value of *x* in this lawn diagram is.

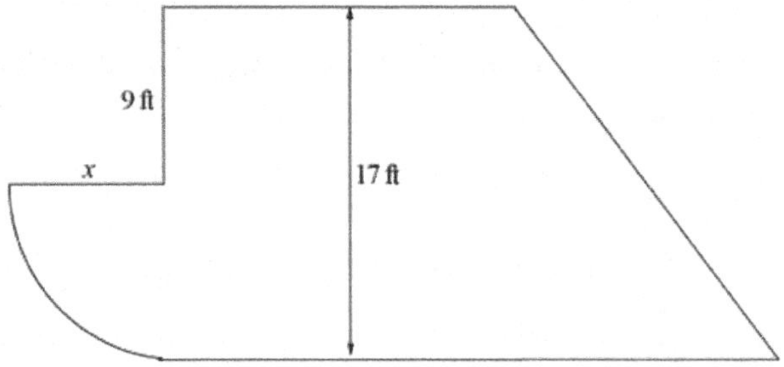

A. 9 ft

B. 8 ft

C. 7 ft

D. 6 ft

E. *x* cannot be determined in this composite shape

• • Answer B.

The radius, *x*, of the circle is (17 ft – 9 ft) = 8 ft.

• The length of x in triangle B is

A. 3
B. 4
C. 5
D. 6
E. Not enough data to find x

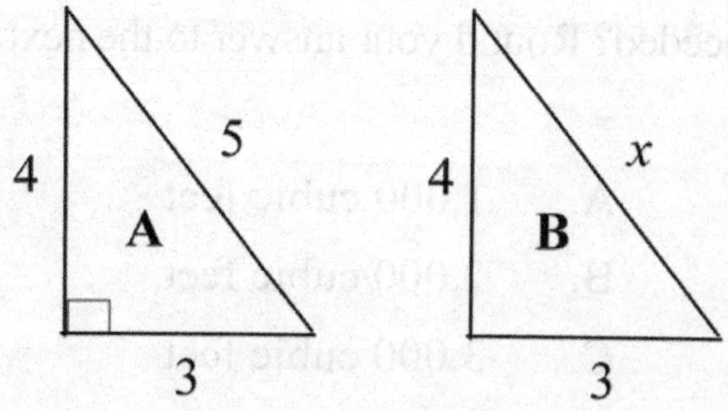

• • Answer E. These are not similar triangles ("A" is a right triangle, "B" is not specified as a right triangle). The angle between the adjacent sides 3 and 4 must be known.

LESSON: if there is no "box" to indicate a right angle, then you cannot assume that the angle is a right angle.

• ♥ •

• Identify the 2- and 3-dimensional figures from the properties listed.

Symbols: A (area); d (diameter); L (length); P (perimeter); r (radius); V (volume); W (width)

A. $C = \pi d$	1. box (parallelepiped)
B. sum of the internal angles is 180°	2. sphere
C. $P = 4L$	3. square
D. V=LWD	4. triangle
E. $V = (4/3)\Pi r^2$	6. circle
F. $A = 2(L + W)$	7. rectangle

A. _6_ B. ____ C. ____ D. ____ E. ____ F. ____

• • Answer: A-6; B-4; C-3; D-1; E-2; F-7

• A negative angle is measured clockwise from a reference side

A. True

B. False

Answer A. True.

Positive angles are always measured counterclockwise from a reference side; a clockwise measure is a negative angle.

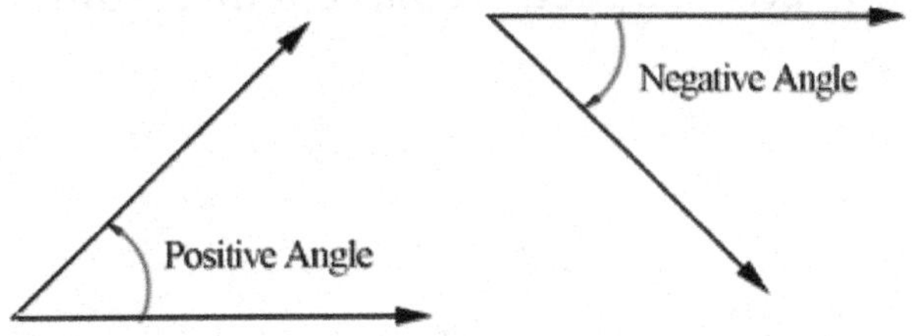

• ♥ •

• The distance to a circle's center from the point of tangency of a line-segment to that circle...

A. is the length of the diameter of the circle

B. is the length of the radius of the circle

C. depends on length of the line-segment

D. is the length of the secant formed by the line-segment

E. None of the above

• • Answer B. The point of tangency is always at right-angles to the radius to that point; therefore, its distance is always the length of the radius.

• The (Cartesian) coordinate plane consists of

A.	an x-axis and a y-axis	[2-dimensions (plane)]
B.	an x-axis, a y-axis, and a z-axis	[3-dimensions]
C.	an x-axis only	[horizontal line]
D.	a y-axis only	[vertical line]
E.	None of the above	

• • Answer A. When a z-axis is added we have a three-dimensional space.

• ♥ •

• Which of the following can be associated with the coordinates (0, 0)?

A. a point

B. a ray

C. a line segment

D. Any of the above

E. None of the above

• • Answer D. (0,0) is a point, and could be the origin of a ray, or one of the ends of a line segment. The coordinates (0, 0) define the origin.

• The statement $\sim(\sim p) \Leftrightarrow p$ is

A. *Modus ponens*
B. Double negative
C. Not a valid logical statement
D. A contradiction
E. A fallacy

Answer B. This is a perfectly valid logical statement

• ♥ •

• Fill in the missing parts of the truth table:

p	q	r	$\sim r$	$q \wedge \sim r$	$p \Rightarrow (q \wedge \sim r)$
T	T	T	*F*	*F*	
T	T	*F*	T	T	
T	*F*	T	*F*	*F*	
T	*F*	*F*	T	*F*	
F	T	T	*F*	*F*	
F	T	*F*	T	T	
F	*F*	T	*F*	*F*	
F	*F*	*F*	T	*F*	

• • Answer:

p	q	r	$\sim r$	$q \wedge \sim r$	$p \Rightarrow (q \wedge \sim r)$
T	T	T	F	F	*F*
T	T	F	T	T	T
T	F	T	F	F	*F*
T	F	F	T	F	*F*
F	T	T	F	F	T
F	T	F	T	T	T
F	F	T	F	F	T
F	F	F	T	F	T

- Which of the following is a logical statement?
  - A. "Hello"
  - B. "I like your hat"
  - C. "Will it rain today"
  - D. All of the above
  - E. None of the above

• • Answer B. I could be lying (I really don't like the hat)

- Complete the missing parts of the table

Symbol	Example	Name
$\wedge$	$p \wedge q$ "p AND q"	________
____	$p \vee q$ "p OR q"	disjunction
$\rightarrow$	________________	conditional
____	$p \wedge q \Rightarrow p \vee q$ "$p \wedge q$ IMPLIES $p \vee q$"	conditional between compound statements
$\leftrightarrow$	$p \leftrightarrow q$ "p if and only if q"	________
$\Leftrightarrow$ or $\equiv$	"p is EQUIVALENT to q" $(p \rightarrow q) \Leftrightarrow (\sim p \rightarrow \sim q)$	________

• •Answer:

Symbol	Example	Name
$\wedge$	$p \wedge q$ "p AND q"	conjunction
$\vee$	$p \vee q$ "p OR q"	disjunction
	$p \rightarrow q$ "if p then q"	conditional
$\Rightarrow$	$p \wedge q \Rightarrow p \vee q$ "$p \wedge q$ IMPLIES $p \vee q$"	conditional between compound statements
$\leftrightarrow$	$p \leftrightarrow q$ "p if and only if q"	biconditional
$\Leftrightarrow$ or $\equiv$	"p is EQUIVALENT to q" $(p \rightarrow q) \Leftrightarrow (\sim p \rightarrow \sim q)$ [contraposition]	equivalent statements

• Complete the truth table below for the disjunction $p \vee q$:

p	q	$p \vee q$
T	T	
T	*F*	
F	T	
F	*F*	

• • The disjunction $p \vee q$ means "p OR q" and so its truth table is:

p	q	$p \vee q$
T	T	T
T	*F*	T
F	T	T
F	*F*	*F*

• ♥ •

• The compound statement "p AND $\sim p$" is a(n)

A. implication

B. contradiction

C. tautology

D. *modus ponens*

E. None of the above.

• • Answer B. "something" and "not something" is clearly a contradiction.

THE MATH TUTOR'S FRIEND

INFORMAL REVIEW OF HIGH-SCHOOL MATH TOPICS FOR TUTORS

- 6 -

• Geometry •

Made in the USA
Monee, IL
02 November 2026

67382583R00050